Lecture Notes in Computer Sci

Edited by G. Goos, J. Hartmanis, and J. va

T0250831

Springer
Berlin
Heidelberg
New York
Barcelona
Hong Kong
London
Milan
Paris
Tokyo

Alessandro Pasetti

Software Frameworks and Embedded Control Systems

 Springer

Series Editors

Gerhard Goos, Karlsruhe University, Germany
Juris Hartmanis, Cornell University, NY, USA
Jan van Leeuwen, Utrecht University, The Netherlands

Author

Alessandro Pasetti
ETH-Zentrum, Institut für Automatik
Physikstraße 3, 8092 Zürich, Switzerland
E-mail: pasetti@pnp-software.com

Dissertation der Universität Konstanz
Tag der mündlichen Prüfung: 8. Juni 2001

Referent: Prof. Dr. Wolfgang Pree
Referent: Prof. Dr. Kai Koskimies

Cataloging-in-Publication Data applied for

Pasetti, Alessandro:
Software frameworks and embedded control systems / Alessandro Pasetti. -
Berlin ; Heidelberg ; New York ; Barcelona ; Hong Kong ; London ; Milan ;
Paris ; Tokyo : Springer, 2002
    (Lecture notes in computer science ; Vol. 2231)
    Zugl.: Konstanz, Univ., Diss., 2001
    ISBN 3-540-43189-6

CR Subject Classification (1998): C.3, D.2.11, D.2.13, J.2

ISSN 0302-9743
ISBN 3-540-43189-6 Springer-Verlag Berlin Heidelberg New York

Springer-Verlag Berlin Heidelberg New York
a member of BertelsmannSpringer Science+Business Media GmbH

http://www.springer.de

© Springer-Verlag Berlin Heidelberg 2002
Printed in Germany

Typesetting: Camera-ready by author, data conversion by Boller Mediendesign
Printed on acid-free paper   SPIN: 10845761   06/3142   5 4 3 2 1 0

Preface

Professionally, the defining experience of my life was a period of nine years as control systems engineer with the European Space Agency (ESA). Given this background, it was perhaps inevitable that when, in 1999, I decided to start a new career as a software engineer I should choose as my area of work software architectures for embedded control systems. As it turned out, this was a lucky choice. After decades of academic neglect, embedded software is now beginning to receive the attention to which its ubiquity in everyday devices and its presence at the heart of many safety-critical systems entitles it. The novelty of the field means that much of the work that needs to be done consists in transferring to embedded systems the technologies and development practices that have become so prevalent in other fields. Following this line of research, I have concentrated on applying object-oriented software frameworks to embedded control systems. Framework technology has enjoyed wide currency for at least five years. Although it has proven its worth as a software reuse technique in many domains, it has been shunned in the embedded world mainly because it tends to be associated with lavish use of CPU and memory, both of which have traditionally been in short supply in embedded systems. Times are changing, though. Hardware advances are expanding the resources available to embedded systems and their adoption of framework technology no longer looks far-fetched or unrealistic.

The first objective of this book is to show how object-oriented software frameworks can be applied to embedded control systems. Software design may not be an art but it is certainly more of a craft than a science and teaching by example is in my view the best way to transfer design experience. I have accordingly chosen a case study as the means to make my point that framework technology can aid the development of embedded control software. The target application of the case study is the attitude and orbit control system of satellites. This domain is broader and less structured than the domains to which frameworks are normally applied. This led me to develop a concept of frameworks that was rather different from that proposed by other authors in that it gives a more prominent, or at least a more explicit, role to domain-specific design patterns. The second objective of the book is to discuss this new view of software frameworks and to present some methodological concepts that I believe can facilitate their development.

This book is therefore written with two audiences in mind: developers of embedded control software will hopefully be inspired by the case study in the second part of the book to apply framework technology to their systems, while researchers on software architectures might be moved to rethink their conceptualization of frameworks by the material presented in first part of the book. The link between first and second part lies in the use of the concepts introduced in the methodological sections of the book to describe the case study framework.

One of the functions of a preface is to offer a path through the book to prospective readers. The key chapters are 3 and 4, introducing the concepts of framelet and implementation case, and chapter 8, giving an overview of the design principles behind the satellite framework. Hurried readers could limit themselves to these three fairly self-contained chapters, which would suffice to give them a rough idea of both the methodological and technological contributions made by the book. The framework concept and the methodological guidelines for framework design are presented in chapters 3 to 7. The case study is covered from chapter 8 to the end of the book. Readers who are not familiar with satellite control systems should read chapter 2 in order to acquire the domain background necessary for an understanding of the case study.

The satellite framework is presented as a set of design patterns that have been specifically tailored to the needs of satellite control systems. Each chapter in the case study part of the book covers one or a small number of related design patterns. These chapters are to a large extent independent of each other and could be read in any order or even in isolation of each other. One way to see this book (or at least its second part) is as a repository of design patterns for embedded control software development.

The appendix at the end of the book contains synoptic tables listing the architectural constructs offered by the satellite framework. The lists are cross-referenced and can serve as an aid to navigate the framework design presented in the book.

This book describes the satellite framework at the "design pattern level". The satellite framework also exists as a publicly available prototype and its full design documentation and code can be downloaded from the project web site[1]. Readers are invited to access this material but they should bear in mind that the code is offered without any guarantee of correctness and that its quality and completeness are those required for a proof-of-concept prototype. They should also be aware that the satellite framework project is a "living project" and its web site is constantly being updated as the framework design and implementation are reviewed and modified. Hence, as time progresses, inconsistencies will inevitably arise with the content of this book.

Having explained what the book offers to its readers, I should write a few words about what it expects of them. The focus is on *object-oriented* frameworks but no effort is made to explain the principles of object-oriented design which are assumed known to the readers. Similarly, familiarity with the design pattern concept is both assumed and necessary. Some background knowledge of framework technology would be useful but is not essential. No prior knowledge of satellite control systems is assumed, all the necessary background being offered in chapter 2. I have made every effort to make the case study as self-contained as possible

---

[1] Its address is subject to change. The site can be found with any internet search engine by searching for "AOCS Framework".

but I suspect that full appreciation of the solutions it offers requires at least a passing acquaintance with embedded control systems. Class diagrams are in standard UML and pseudo-code examples are written using C++ syntax with the addition of the interface construct from Java which I find simply too useful to ignore.

Finally, I wish to thank Wolfgang Pree and Kai Koskimies who reviewed the first draft of this book and Jean-Loup Terraillon, Ton van Overbeek, Richard Creasey, and David Soo who, in various capacities, supported the satellite framework project.

August 2001   Alessandro Pasetti

Contents

1 Introduction and Context

There are so many books in print that the first duty of an author must be to explain why a new one is needed. The present book is about the application of software framework technology to embedded control systems and its core is the description of a case study of a full object-oriented framework for a particular – but representative – class of embedded control systems. Its justification therefore requires arguing the case in favour of the distinctiveness of these systems. This is done in the next section with reference to embedded systems in general since their claims to distinctiveness encompass and subsume those of the narrower category of control systems. The following sections instead outline the long-term programme of research within which the results presented here were obtained. The material in the book only covers a small part of this programme (which, at the time of writing, is still on-going) but understanding the latter helps understand the motivation and intention behind the former. The chapter closes with a formulation of the objectives of the book and a summary of its contributions.

1.1 The Embedded Software Problem

It is probably true that, as many believe, nothing of any significance has been invented in computer science since the late sixties. Possibly in order to compensate for this dearth of new ideas, software engineers have proven exceptionally adept at introducing new words for old concepts and at attaching new meanings to old words. Whether for this or for other reasons, their discipline is plagued with terminological confusion. It is therefore fitting that this book should begin with a definition (with many more to follow in later chapters). In accordance with common usage, the expression *embedded system* is used to designate a computer system hidden inside a product other than a computer [95]. This definition can be turned inside out to say that an embedded system is a computer system that is used to control the product that contains it. Embedded systems are therefore the means by which the power of software is harnessed to make the behaviour of non-computer devices – from washing machines to satellites – more flexible and more controllable. The *embedded software* is the software that runs on an embedded computer and that is the ultimate source of this flexibility and controllability.

In what sense is embedded software different from other types of software and why does it deserve dedicated treatment? Answering this kind of question inevitably exposes an author to the twin dangers of overgeneralization and oversimpli-

A. Pasetti: Software Frameworks, LNCS 2231, pp. 1-16, 2002.
© Springer-Verlag Berlin Heidelberg 2002

fication. These dangers are especially acute in the present case because of the exceptional diversity of embedded systems. Still, even where there is diversity there can be average trends and general patterns and it is to them that the discussion that follows applies.

At first sight, the differences between embedded and non-embedded software have two aspects, functional and technological. Functionally and in a negative sense – looking at what embedded systems are *not* or do *not* have – embedded software is characterized by the complete lack, or at least sharply reduced importance, of the user interface. This already sets it apart from other software since the user interface in conventional systems accounts on the average for 50% of the total development and coding efforts and, when a graphical user interface is used, this proportion is even higher and can reach 90% [28].

In a positive sense – looking at functional features that embedded systems have and other systems lack – embedded software is primarily distinguished by its direct interaction with hardware peripherals. In the case of embedded control systems, for instance, the software always has to manage a set of sensors from which measurements about the state of the variables under control are acquired and a set of actuators to which command signals are periodically sent. The need to respond to and control external hardware and external processes often results in timing requirements being imposed on embedded software. Embedded systems must then be built to ensure certain minimal response times or to guarantee a certain minimal throughput. Meeting such requirements can have a decisive impact on the architecture of the software and timing aspects are invariably among the design drivers of embedded software. Additionally, embedded systems are by their very nature designed to operate outside direct human supervision. Indeed, their task is often to replace or complement human supervision of a certain process or device. Standalone operation requires a high degree of autonomy and reliability to be built into embedded software.

Functional differences, though important, are not those that most stand out in a comparison between embedded and non-embedded software. The truly remarkable difference between the two is technological and relates to the vast gap that can be observed between the level of technology that is now common in general purpose computer systems and the level of technology prevalent in embedded applications. When seen from the point of view of mainstream computer science, embedded software projects often appear to be taking place in a time warp: the language of choice remains C (with generous sprinklings of assembler), the dominant architectural paradigm is only now shifting from procedural to object-based, software engineering tools are often ignored.

Technological backwardness is the norm even in fields that are perceived as "high-tech". In 1996, the author of this book was asked to take part in the review of a control system for a large satellite being developed by a major aerospace company for the European Space Agency and was astonished to find that its software was entirely written in assembler (using a modular approach – as one of the project engineers proudly declared). For another example, we can turn to the car industry. As recently as 1997, the engine control software of a very popular line of middle-sized cars consisted of about 300.000 lines of C code that was poorly structured because of its assembler heritage, and that, after years of changes to fulfill ever changing requirements, had become virtually unmaintainable [69].

Technological backwardness can be confirmed by inspection of some of the recent reference books on software for embedded systems such as [95, 46, 59, 7]. Although all four texts mention C++ or object-orientation (sometimes even in their titles) and give due emphasis to the virtues of the latter, the paradigm that underlies most of the material they present is the traditional object-based or modular one. The same impression can be derived from the "Embedded Systems Magazine", a prime reference for embedded engineers. Advanced software topics – object-orientation, component technology, real-time Java and others – are sometimes broached but their adoption in practice remains controversial and is clearly limited to a small minority of existing systems (for a typical example of this attitude, see the cover article of the issue of August 1999 with the eloquent title of: "Nuts to OOP!").

It might be expected that research would tend to redress the imbalance between embedded and other software. In reality this is far from being the case. Discipline folklore has it that while 99% of processors that are sold in a given year are used in embedded devices, only 1% of computer scientists are working on embedded software. These figures are difficult to verify but there is no disputing that the paradox they point to – the overwhelming majority of software research being focused on a narrow subset of all computer systems – is real.

Lack of interest on the part of software scientists can be largely explained by the limited resources available to embedded computer systems which typically have memories restricted to kilobytes and processor architectures of the CISC, 8- or 16-bit kind. Application memory and CPU margins are correspondingly tight and seldom leave any scope for applying the advanced software technologies that are the staple of academic publications and that are so widespread in desktop applications.

To some extent, this situation is self-perpetuating. Lack of research in embedded software leads to lack of innovation and to software systems that are comparatively small and simple. Since embedded devices are often mass-produced, profit margins are largely determined by hardware costs and manufacturers have an incentive to continue using the minimalist processor configurations that have stifled research and innovation in the first place. Perhaps more subtly, lack of interest in embedded software on the part of the research establishment has turned the subject into one of computer science's poor relations which hardly encourages software scientists to enter the field and attempt to redress the imbalances described above. The web of causal interrelationships between technical, economic and socio-cultural factors that contribute to keep embedded and other software apart is sketched in figure 1.1.

Recent developments, however, may herald a reversal of fortune for embedded software and its practitioners. On the one hand, the continuing decline in hardware costs is bringing more and more processing power within the budgetary envelope of embedded software projects while, on the other hand, ever expanding consumer expectations are putting increasing demands on the software, demands that can only be accommodated by resorting to the same kind of technology prevalent in the non-embedded world.

If disparity in the level of hardware resources were the only – or the main – factor differentiating embedded from other software, then there should be no reason to treat embedded software as "special": as the resources available to embedded developers are brought in line with those available to their desktop colleagues, one should see embedded and non-embedded software naturally converge towards the same technological level and towards the same design and implementation practices. This is unlikely to happen and this is a symptom of a more profound, albeit often overlooked, difference between much embedded and much non-embedded software.

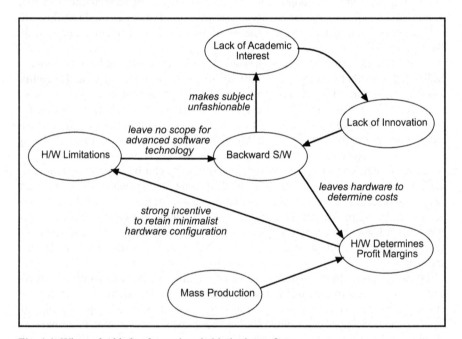

Fig. 1.1. Why embedded software lags behind other software

Software design is always a difficult undertaking but it is especially so when the persons doing the design have a poor understanding of the target application. This problem occurs everywhere but it is more severe in the embedded world. After all, the definition of embedded software given at the beginning can be paraphrased to read: "an embedded software is a piece software for a system other than a computer". The software for an embedded system consequently has a somewhat accidental character, it serves to embody functions that find their origin and motivation in the non-software aspects of the system. Software development for embedded applications is thus often characterized by the divorce between the roles of the software specialist – the person responsible for the software development – and the application specialist – the person responsible for specifying the application. The gap between the two is normally bridged by some kind of formal description of the application to be developed such as user requirements or use cases. The difficulty of exhaustively and unambiguously capturing the specification of a complex system means that, almost inevitably, as a project progresses, the interface between the software and application specialists becomes a source of personal attrition, schedule delays, cost overruns and specification misunderstandings. This problem is all the more serious because more and more of the functionalities of an embedded system are concentrated in its software and the split of work between the software and domain specialist therefore results in a situation where the heart of the application – its software – is designed, developed and tested by persons who are *not* application specialists.

It is worth to make this point by means of an example. First, consider the case of a software engineer who is asked to develop the software for, say, a new word processing application. As an input, he would probably be given a set of user requirements. Since he is familiar with word processors (because he has used them, because he has studied them at university, because he has read about them, etc) it is very likely that he would quickly understand them and would be able to assess their quality by, for instance, spotting mistakes, gaps, and ambiguities. Additionally, having a good "feeling" for word processing applications, he would be able to judge whether the requirements he has received are implementable and would quickly identify those that are critical and need special attention.

Suppose now that the same software engineer were asked to write the software for, say, a GPS receiver (a typical embedded application). GPS receivers are likely to be outside his typical domain of experience and knowledge and he would have to rely completely on the formal description of the GPS software that he receives from his customer. Firstly, he might have some trouble understanding it because he is not familiar with the vocabulary of that domain (whereas he was perfectly at home with the domain vocabulary of word processors). Secondly, he would not be in a very good position to assess the quality of the inputs he receives and to identify potential problems due to inconsistencies and misunderstandings. These problems would eventually emerge but they would emerge only during application implementation or, even worse, during testing.

It is obvious that, other things being equal, our software engineer would make a much worse job in the second than in the first case and that the cause of this is the fact that, in the second case, he lacks domain understanding. Our contention is that the scenario described in the second example typifies embedded systems in general and embedded control systems in particular.

Finally, and for completeness, it must be mentioned that in practice a common means for the application engineers to by-pass their software colleagues and avoid the problems and costs that arise at the interface between the two disciplines is for them to write their own code directly. This, however, is not a viable solution in any but the simplest cases: software development requires specialized skills that normally only specialists possess. The solution to the difficulties discussed above lies in optimizing the role of each of the two specialists in the application development process, not in ignoring the need for their cooperation.

1.2 Empowering Application Specialists

Three differences between embedded and other software have been identified. From a functional point of view, embedded software is characterized by tighter coupling with hardware, more stringent timing requirements, a reduced user interface and greater autonomy needs. These, however, are differences of degree rather than kind and would not by themselves justify treating embedded software as "special". Secondly, embedded systems are constrained by limited hardware resources which accounts for their technological backwardness. This difference is expected to wane in importance as more powerful processors and larger memory become cheaper and find their way into embedded devices. Finally, there is the cultural factor due to the lack of a common background between the application and software specialists. We believe that this is at the root of the distinctiveness of embedded software and we think that this is the problem that research in embedded software should address.

Arguably the most straightforward way to remedy it is through the development of better techniques to capture application requirements. An entire field of research has grown around this solution but has so far failed to produce any breakthroughs. An alternative and more radical way to address the same problem is to remove the interface between the application and the software engineer from within a project. This can be done by providing tools that empower the application specialists to directly develop the software for their system without the need for the intermediary services of a team of software engineers. The role of the latter then becomes focused on the development of the tools themselves, not on the development of the applications.

The two ways of developing application software are contrasted in figure 1.2. The flow of activities at the left shows the traditional approach where the application expert hands over a set of formal requirements to a software expert who uses them to develop the application software. The right hand side of the figure instead shows the alternative approach where the software specialist cooperates with the application specialist to develop domain-specific tools that are intended to allow application engineers without specialized software expertise to directly build their own software.

In order to be more specific, and out of consideration to the general subject of this book, the discussion that follows is restricted to embedded control systems but its conclusions would apply, *mutatis mutandis,* to many other embedded domains. Embedded control systems are interesting in this respect because they are one field where the approach advocated here already finds concrete application. Control engineers reading this book are likely to be familiar with tools like MatLab [65] or Xmath[2] [100] which offer them a sophisticated environment where they can rapidly design a control system using the abstractions of their discipline (transfer functions, state estimators, PID controllers, etc). After the design has been validated by means of simulations, the environment can automatically generate

---

[2] Matlab and Xmath are registered trademarks.

a software module that implements it and that is ready for downloading to a target processor. Many control specialists believe that such environments hold the promise to solve the "software problem" by removing the need for an explicit software development phase. This claim has been tested on real systems [61, 32] but, as will be argued in the next chapter, we believe that such tools are inadequate to completely replace the software engineering activities. The philosophy that underlies their usage, however, is correct since they shift the role of the software engineer from the traditional one of developing the software on behalf of the application specialist, to that of developing tools that empower the application specialists to directly assemble their own software.

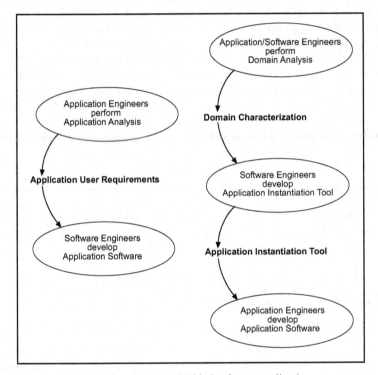

Fig. 1.2. Two approaches to developing embedded software applications

The main shortcoming of environments *à la* Xmath/Matlab is that, although they do an exceptionally good job of modelling and implementing in code the control algorithms, this only represent a small subset of the software in a complex control application (in a satellite control system it would be about 20-30%). They therefore need to be complemented with techniques to handle the remainder of the embedded control application (covering interactions with an external operator, failure detection and failure recovery tasks, management of external units, etc). This is the point where software frameworks must be introduced.

Software frameworks are a software reuse technology that fosters the reuse of entire architectures (as opposed to code fragments) for applications within a certain domain. The framework defines a generic architecture that can be adapted to rapidly instantiate a particular application within that domain. A software framework offers ready-to-use components that encapsulate recurring abstractions within the domain. Applications are built by combining plug-compatible blocks that represent concepts, functions and objects that are useful within the application domain. There is an important conceptual similarity with tools like Xmath/Matlab which owe their popularity to the fact that they allow the control engineers to define a control system in terms of high-level abstractions that are meaningful to them as control engineers. A software framework does a similar service for the developers of applications in a certain domain because it lets them build an application by manipulating components that represent abstractions that are directly relevant to their domain of interest. To use an example from the GUI field – where the framework approach has been especially successful – it easier to build a GUI application by starting from configurable components that represent menus and windows rather than by having to deal with subroutines that can simply draw lines and position the cursor on the screen. A framework for embedded control systems can therefore be seen as a natural complement for Xmath/Matlab tools if it can be made to offer encapsulations of those abstractions that are specific to the embedded control domain and which the latter do not cover.

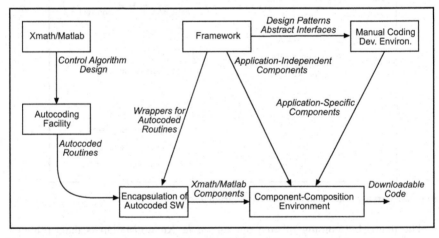

Fig. 1.3. Development environment for embedded control applications

The work presented in this book is part of a wider programme of research that tries to leverage the specific advantage of Xmath/Matlab and software frameworks to realize an environment where control engineers can develop new applications with a minimum amount of manual coding. The structure of this environment is shown in figure 1.3.

The first actor in the application construction process is a Matlab/Xmath tool which allows the control engineers to define and validate by simulation the control algorithms. Its autocoding facility is used to generate the code encapsulating the algorithms thus defined. These algorithms are finally packaged as components using pre-defined wrappers. The non-algorithmic part of the application – its control logic and the data handling parts – are defined in a separate and dedicated environment built on a software framework that provides a configurable architectural skeleton for the application and a set of default components representing standard implementations of common domain functionalities. Inevitably, some manual development of *ad hoc* components will be required but this will be aided – and constrained – by the need to adhere to the design patterns and abstract interfaces exported by the framework. Compliance with the framework design patterns and interfaces facilitates the development of the new components and it preserves the architectural integrity and high-level uniformity of the application by ensuring that similar problems in different parts of the application receive similar solutions.

Application development therefore begins by collecting a number of components, some derived from the Xmath/Matlab environment, others obtained by suitably configuring pre-defined framework components, and still others (hopefully a small minority) hand-coded to implement in an application-specific manner selected framework design patterns. These components then need to be composed to form a complete application. This is essentially the process of framework instantiation. This process is typically done by manually writing the code to glue together the components but could be automated (as is already the case in the field of GUI applications). Ideally, a tool should be provided where the user interacts with a visual environment that offers a palette of predefined components and allows them to be configured and connected visually with the glue code being automatically generated to reflect the choices made by the developer. This would complete the analogy between software frameworks and Xmath/Matlab since an important part of the latter's appeal is the ease of use of their graphical user interface.

It is worth stressing a difference between Xmath/Matlab and software frameworks. The former typically have a broader application range than the latter. They are targeted at control systems in general whereas frameworks normally have a more restricted field of application. The framework presented in this book, for instance, is aimed at the attitude and orbit control system of satellites. The applicability of an instantiation environment as shown in figure 1.3 is therefore given by the intersection of the target domain of the Xmath/Matlab environment with that of the software frameworks. Normally, the latter will be the most constraining of the two.

1.3 The Component Software Challenge

We have argued that some categories of embedded control systems – we used the example of satellite control systems – are becoming mature for the application of modern software techniques. Out of the many possible directions in which their technical development could proceed, this book concentrates on object-oriented software frameworks. There are at least two perspectives in which this choice can be put into a broader context. In one sense, software frameworks can be seen as an enabling technology that will empower application specialists to take full control of the design and development of the software for their systems. This perspective was discussed in the previous section. The present section instead considers the view of object-oriented frameworks as a stepping stone towards the adoption of a component approach to the development of satellite control software [73].

The problem that software component technology [98] aims to solve is that of software reuse. In some fields – such as desktop applications – this technology has by now become well established and has led to important gains in productivity with consequent reductions in cost and development times. These gains are made possible because software systems come to be built as hardware systems are: as assemblies of pre-defined, off-the-shelf items. By contrast, software reuse in embedded systems in general and in control systems in particular remains low and scattered. The roots of this low degree of reuse probably lie in the way the software development process is structured. Typically, this process begins with the definition of user requirements. Requirements are normally formulated under the tacit assumption that all the software will be developed anew. Consequently, they are narrowly targeted at a specific mission and result in a monolithic piece of code that, while highly optimized for the application at hand, has to be developed entirely from scratch and cannot be reused in other missions.

This approach is in sharp contrast to that adopted on the hardware side. Here, designers begin by surveying the market for available components and then specify their system in terms of these components. Widely accepted standards (on bus interfaces, on electrical interfaces, on mechanical interfaces, etc.) allow components from different suppliers to be combined. The resulting system is perhaps not optimal in terms of mass or power consumption (the characteristics of standard components seldom match perfectly the requirements of a specific project) but it is certainly cheaper and faster to assemble than if it were developed from scratch. This difference between software and hardware development processes is not an ineluctable consequence of a "special" nature of software systems. In fact, in some (non-embedded) fields it is disappearing thanks to the introduction of *software component standards*. The best known examples of such standards are Microsoft's (D)COM, OMG's CORBA, Sun's JavaBeans. Their function is analogous to that of hardware standards: they define interfaces and communication protocols that allow pieces of code developed as self-contained units, and potentially supplied by different vendors, to cooperate.

It is important to emphasize that software components go beyond traditional subroutine or class libraries. The latter do not qualify as components because their interfaces are idiosyncratic and because calling and called code must normally be written in the same language and compiled with the same compiler. Component standards by contrast specify industry-wide interfaces that allow independent units of code to interoperate across language, compiler and even platform barriers. This is a crucial difference since it might lead to the emergence of a market for third-party software components not dissimilar to the markets for hardware items. As a consequence, complex software systems might come to be developed very much as hardware systems are developed: as collections of cooperating pre-defined components.

Components are not a new concept in the embedded control field. Real-Time Operating Systems (RTOS) are examples of components with non-standardized interfaces. After their introduction, engineers no longer had to develop operating systems for their applications. Instead, they could choose a commercial product, configure it for their purposes, and integrate it with their own code. The challenge is to extend this approach to other parts of embedded applications – sensor and actuator management, reconfiguration management, control algorithm implementation, failure detection, etc – ultimately arriving at applications that consist mainly of assemblies of separately procured components. The full realization of this vision lies well into the future. How far exactly into the future obviously depends on which category of applications one is considering. The case study later in the book concerns satellite control systems and, for this class of applications, table 1 maps the path from the present to the ideal future of a component-based software by breaking it up into six intermediate milestones. Similar tables could be built for other categories of software (and the results are likely to be similar).

Table 1. Technical challenges on the way to component-based software in space

Technical Challenge	Description
Hardware Challenge	Component technology requires greater hardware resources than traditionally available on board satellites.
Software Challenge	Component technology requires use of OO language whereas satellite software is traditionally written in C or Ada83.
Methodological Challenge	Design methodologies for space software are targeted at monolithic applications of procedural or object-based kind. New methodologies aimed at OO and component-based systems are needed.
Architectural Challenge	Satellite software is developed as a monolithic mission-specific application. A component approach requires it to be re-architectured as a set of smaller, loosely coupled units.
Real-Time Challenge	Satellite systems have real-time constraints. Existing component infrastructures must be modified to be suitable for real-time environments.
Robustness Challenge	Satellites have demanding reliability requirements. Component software will increase complexity and hence likelihood of failures. New failure handling strategies must be developed.

Of the six technical challenges listed in the table, the first two can be considered to have been met. As already mentioned, the introduction of SPARC processors for use in space has solved the hardware problem. The software problem in its narrow sense of availability of object-oriented tools and languages has also been largely solved because development systems are now on the market that allow object-oriented languages (Ada95 or C++) to be coupled to commercial real-time operating systems for these space-qualified processors. At the other end of the journey towards component-based satellite software, the last technical challenge in the above list must be considered to be widely open. Failure handling is an extremely difficult issue even in present systems and it is unclear how current approaches should be modified to cope with the greater complexity introduced by the use of components. Currently, failures arising from design flaws are regarded in space systems as catastrophic events catered for only through a "safe mode" which merely guarantees the physical survival of the satellite but does not allow it to continue its mission. Greater software complexity may ultimately require functionality levels that are intermediate between "fully operational" and "safe mode" but how this can be done in practice is as yet unclear.

Most current research work concentrates on the intermediate challenges 3 to 5. The present book in particular makes a contribution to the methodological and the architectural challenges. The methodological problem has in fact a wider relevance extending beyond satellite systems. The introduction of the new software technologies of the 90's has proceeded much faster than the ability of the wide user community to master them. One contribution of this book lies in the proposal of methodological concepts for object-oriented frameworks. These concepts are of course targeted at framework technology but because of the stress they place on design by interfaces and because they conceptualize an applications as a set of plug-compatible component-like objects they represents a step in the direction towards a methodology for component-based systems.

The architectural problem is addressed in a more decisive manner since the heart of the book is the definition of a generic architecture for satellite control systems based on a set of independent components encapsulating functionalities that are potentially reusable across missions. Reusability was achieved by splitting the *management* of a functionality (which is mission-independent) from its *implementation* (which remains mission-specific). As already noted, RTOS's are one successful example of component reuse in the embedded field. The case study framework proposed here has deliberately tried to replicate their success by extracting from a satellite control software OS-like functionalities that can be encapsulated in reusable components.

The building blocks in the architecture proposed by the case study framework are components only in the sense of being self-contained, binary deployable entities that implement well-defined interfaces. Realizing the vision outlined at the beginning of this section of a truly component-based satellite control system would require their being implemented on a component infrastructure *à la* CORBA or (D)COM. This would allow interoperability across compiler, language, process and even processor boundaries with obvious positive impacts on software standardization and software reuse. One reason why porting to a component infrastructure cannot yet be done is that satellite control systems are hard real-time systems and the real-time component challenge (point 5 in the table) remains still unmet. This may soon change since much work is in progress to make component

standards real-time: an ORG working group has proposed a specification for a real-time CORBA [70, 87], and Sun is considering a real-time extension of Java
[86], possibly the first step towards turning JavaBeans into suitable vehicles for
real-time components. In the meantime, the TAO project [45] offers an interesting
alternative approach. Its designers, rather than waiting for a new specification,
have provided an implementation of the current CORBA standard that guarantees
that calls across component boundaries preserve priority levels and that the overhead in servicing a call request is statically predictable. This makes the ensuing
system amenable to the same static schedulability analyses as are used in traditional real-time systems.

Table 2 summarizes the status of the six technical challenges on the way to the
goal of fully component-based satellite software. This book is obviously far from
solving them all but, as discussed above, it gives an important contribution to solving both the methodological and architectural problems at least for one category
of embedded control software and probably – because of the representativity of
the selected application domain – for embedded control systems in general.

Table 2. Status of technical challenges for component-based space software

Technical Challenge	Current Status
Hardware Challenge	Addressed by introduction of space-qualified SPARC processors.
Software Challenge	Addressed by introduction of object-oriented development systems for space-qualified SPARC.
Methodological Challenge	In the process of being addressed. Contribution from this book through definition of design methodology for object-oriented frameworks.
Architectural Challenge	In the process of being addressed. Contribution from this book through definition of component-based architecture for the AOCS.
Real-Time Challenge	In the process of being addressed.
Robustness Challenge	Still open.

1.4 Objectives and Contributions

Embedded systems are too diverse to be treated in any detail as one single entity.
Making a practical contribution demands a narrower focus and in this book we
have chosen to concentrate on embedded control systems and in particular on a
case study for the Attitude and Orbit Control System (AOCS) of satellites. The
chief contribution of the book is the description of a full object-oriented software
framework for this class of control systems. With reference to its target domain, it
will be called the "AOCS Framework". This framework was originally developed
in a project partially funded under research contract Estec/13776/99/NL/MV for

the European Space Agency (ESA) that saw it as a first foray into the area of software frameworks for space systems[3].

Design of the AOCS Framework required a complete reconceptualisation of its domain. This entailed, on the one hand, the identification of the invariant domain functionalities and their encapsulation in abstract interfaces and concrete components, and, on the other hand, the identification of the points of behaviour adaptation of individual AOCS applications and the definition of domain-specific design patterns to model them. To our best knowledge, the AOCS Framework represents the first attempt to apply object-oriented design techniques to satellite on-board software and one of the first to apply them to a complex embedded system. Since satellite control systems are representative of embedded control systems in general, many AOCS framework components and design patterns would be applicable to other categories of control applications and the architectural solutions described here will therefore be of interest outside the satellite community.

Software frameworks were presented above as an enabling tool for application specialists. In the long run, it is hoped to integrate the AOCS Framework in an environment as depicted in figure 1.3. This however does not mean that the framework is not in the meantime useful by itself. Software frameworks were introduced as vehicles to increase software reuse and there is no doubting that embedded control systems can benefit from greater reuse. In the satellite field, reuse, when it takes place at all, seldom concerns more than isolated code fragments that are carried over across projects when the same company – or probably the same individuals – happen to be in charge of two related projects. There is much anecdotic evidence to suggest that similar situations prevail in other fields. Development of a software framework would force a company to plan for reuse and to invest in reusable assets. We hope that the framework presented here can act as an inspiration and a blueprint for the construction of such assets for embedded control systems and can provide a repository of design patterns for this domain.

The AOCS Framework is also interesting because its domain – as many other control domains – must normally satisfy hard real-time requirements. Framework technology has not been applied often to this type of applications. The AOCS Framework by contrast explicitly tackles the real-time aspects and has built-in provisions to ensure that applications instantiated from it can be statically analyzed for their timing properties.

A further contribution stems from the particular approach that was followed to design the AOCS Framework which was based on an analogy between frameworks and operating systems leading to an original view of frameworks as domain-specific extensions to the operating system. From a theoretical perspective, this view is useful for the insight it offers on the framework concept. From a practical perspective, its value is threefold: conceptualization of the AOCS Framework as a domain-specific extension of the operating system helped suggest design solutions during the framework development; it now facilitates the description of its archi-

---

[3] The views expressed in this book are those of its author only. They neither commit ESA not represent official ESA thinking.

tecture; and in the future it will make framework upgrades easier. This structure of the framework offered another advantage. A software framework builds a degree of adaptability into an application that has an inevitable impact on its performance and memory requirements. Some of these overheads are related to the use of object-orientation (e.g. timing overheads due to dynamic binding of methods, memory overheads due to storage of virtual tables, etc). These are well-known, have been analyzed elsewhere and, to some extent, have been quantified [34, 48]. Overheads specifically tied to the adoption of a framework approach are less well known The structure of the AOCS Framework as an extension of the operating system makes their measurement possible in the same way that the overheads introduced by an operating system are measured.

The initial thrust of the AOCS Framework project was purely technological. However, very soon, it became clear that methodological issues would have to be considered too. A new conceptualization of software frameworks had to be elaborated which, unlike most of those found in the technical literature, would be capable of capturing the commonalities of a domain as complex as that of the AOCS and of supporting reuse at a higher level of abstraction than is usually done. This new view of software frameworks represents the first methodological contribution of this book. Its peculiarity lies in the role that it assigns to domain-specific design patterns as constituent blocks of the framework and in the explicit distinction it makes between *design* and *architecture*.

Secondly, the complexity of AOCS systems made development of a framework as a monolithic entity impossible. The design problem, to be manageable, had to be broken up into smaller problems. This led to the introduction of methodological tools – *framelets* and *implementation cases* – to tackle framework design complexity. Their use is integrated within a design procedure for software frameworks that formalizes the approach we followed in developing the AOCS Framework. This design procedure and the conceptual tools that underlie it represent a methodological contribution that is interesting in its own right as it might be applicable to other framework development processes and that we hope will earn this book an interested audience among software engineers as well as control engineers.

In summary, this book is primarily concerned with presenting the results – both technological and methodological – of the AOCS Framework project. The previous three sections have presented three angles from which these results can be viewed. Each angle provides a different perspective and points to a different problem area to which it can contribute. These perspectives are schematically brought together in figure 1.4. As shown in the figure, the project acts upon its context (the three bubbles in the figure) by providing partial solutions (the arrows in the figure) to the problems that this context defines. This figure highlights the contributions that the AOCS Framework project makes to its context but the context of the project also defines the directions in which its results can be extended. Thus, the figure can be recast as in figure 1.5 to show how each aspect of the project's context points to different extensions of the AOCS Framework project as represented by the arrows in the figure.

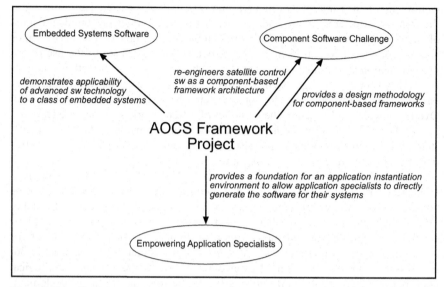

Fig. 1.4. Contributions of the AOCS Framework project to its context

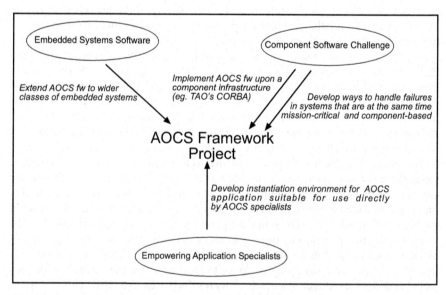

Fig. 1.5. Possible directions of extensions for the AOCS Framework

2 Attitude and Orbit Control Systems (AOCS)

When discussing software engineering topics, it is only too easy (and too temp-ting) to remain on a purely theoretical plane and speculate at length about abstru-senesses of dubious practical relevance. To protect against this danger, this book concentrates on a concrete case study consisting in the design of an object-oriented framework for Attitude and Orbit Control System or AOCS of satellites. Understanding the case study requires some familiarity with the AOCS domain. This chapter presents the necessary background. There is much variation across satellites in the way attitude and orbit control is performed. The slant of the de-scription here is towards the AOCS subsystem as it is found in scientific satellites developed by ESA. Those who are interested in a more comprehensive and more detailed view are directed to reference [101] which is the standard reference in this field. Reference [62] is also relevant in that it covers the broader satellite mission context.

2.1 AOCS Systems

In presenting satellite control systems, it is well to start from the overall satellite context. The left side of figure 2.1 shows a "black box" view of a satellite system. The satellite is controlled from a ground station through a radiofrequency link. Depending on the mission, the link with the ground station may be continuous or it may extend only over a section of the satellite orbit. The satellite then has two external interfaces:

– *Telecommands*: commands sent from the ground to the satellite;
– *Telemetry*: data sent by the satellite to the ground.

Telecommands are commands sent by the ground station to the satellite. They determine the behaviour of the satellite and can sometimes override internal deci-sions taken by the on-board software. Telecommands are normally sent as asyn-chronous data packets. Telemetry data are sent by the satellite to the ground in or-der to, among other things, give it information about the general status of the satellite. Telemetry data are sent in packets. In the past, the sequence, size and content of packets was fixed. In more recent systems, the ground can select which packets to acquire and how often they should be provided. Each satellite subsy-stem may contribute its own telemetry. Thus, for instance, in the telemetry flow, there is a stream of AOCS data that are generated by the AOCS.

A. Pasetti: Software Frameworks, LNCS 2231, pp. 17-28, 2002.
© Springer-Verlag Berlin Heidelberg 2002

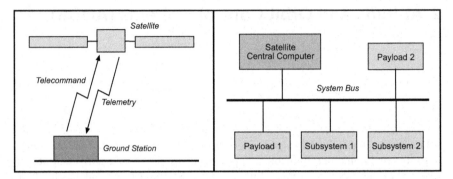

Fig. 2.1. (Left) High-level view of a satellite; (Right) Internal structure of a satellite

The right side of figure 2.1 shows a typical internal system architecture of a satellite. The satellite system is built around a *system bus*. A satellite central computer acts as bus master. The clients on the system bus are the *satellite subsystems* and the payloads. Satellite subsystems often have their own computer and may even have an internal bus. The AOCS is normally one satellite subsystem. The data travelling over the system bus are telecommands and telemetry. Telecommands are distributed by the central computer to the subsystems and payloads, while telemetry is sent by the subsystem and payload to the central satellite computer. All normal communications between central computer and subsystems and payloads take place over the system bus. However, other links exist for use in special situations such as initialization, power-up, failures or other non-nominal tasks.

The AOCS subsystem is one of the subsystems of a satellite. It is often the most complex of the satellite subsystems. Its function is to control the attitude and the orbit of the satellite. This is a critical function because without attitude or orbit control the satellite will gradually – sometimes in a matter of seconds, sometimes after several days – become unable to fulfill its mission objectives.

Figure 2.2 shows the conceptual structure of the AOCS. The figure assumes an AOCS with a dedicated computer separate from the central satellite computer. In an alternative configuration, the AOCS software is hosted on the central computer. The external interfaces of the AOCS are the telecommands it receives from the satellite central computer and the telemetry it forwards to it over the system bus. Other commands for special contingencies (such as for instance forcing the AOCS into safe mode), may be routed over dedicated command lines. Internally, the AOCS periodically acquires measurements (on the satellite attitude and orbital position) from a set of sensors and uses them to compute control signals that are sent to a set of actuators.

Because of its complexity, the AOCS is often organized around a dedicated AOCS bus over which communications between the AOCS computer and the AOCS sensors and actuators take place. In other cases, the satellite system bus doubles as AOCS bus.

Because of their mission-criticality, AOCS systems are normally required to be single-fault-tolerant meaning that they must be able to guarantee full performance in the presence of any single fault. This objective is achieved by making the AOCS entirely redundant. This implies that there are two AOCS computers (prime and redundant computer), two AOCS buses (prime and redundant bus), two sets of AOCS sensors and actuators (prime and redundant units). Sometimes, switchover can only be performed from the prime AOCS to the redundant AOCS. In other case, switchovers at the unit or bus level are possible. Redundancy switchovers can sometimes be triggered autonomously by the AOCS (or by other satellite subsystems) whereas in other cases they are under the exclusive control of the ground station.

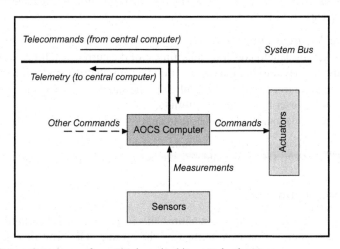

Fig. 2.2. Internal structure of an attitude and orbit control subsystem

It is interesting to note that the software is normally not redunded. This is because software can only fail if its design is faulty and current satellite systems normally adhere that only errors due to accidental malfunctions should be protected against. Typical examples are hardware failures in peripheral units. The rationale behind this policy is a wish to minimize on-board complexity which dictates that design errors should be prevented through appropriate design verification procedures on ground rather than through on-board checks.

2.1.1 AOCS Functions

The objective of the AOCS system is the preservation of the satellite attitude and the control of its orbit. At any given time, a nominal attitude is defined for the satellite. This is the attitude that the satellite should ideally maintain. For instance, in the case of a geostationary telecommunication satellite, the nominal attitude has the satellite pointed towards the earth. In another example, in the case of an astronomical observatory, the nominal attitude has the boresight of the telescope pointed at the astronomical object that it is desired to study. The nominal attitude may be either telecommanded by the ground or it may be generated internally by the

AOCS software. In order to achieve its twin objectives of controlling the satellite attitude and orbit, the AOCS implements several main functionalities as described in the subsections that follow.

Attitude Control Function

The attitude is controlled in closed loop and autonomously on-board the satellite. The satellite attitude is controlled by applying torques to the satellite. Torques are applied by *attitude actuators*. The attitude control function is executed cyclically and its steps are:

- Acquire data from attitude sensors;
- Process sensor data to construct an estimate of the current satellite attitude;
- Compute the deviation of the current attitude from the nominal attitude;
- Compute a control torque to be applied to the satellite to bring the satellite actual attitude closer to its nominal attitude;
- Send commands to attitude actuators to apply the control torque to the satellite.

The typical frequency for the attitude control cycle is about 2 Hz but higher frequencies up to 20 Hz are possible.

Orbit Control Function

A nominal orbit is defined for the satellite. This is the orbit that the satellite should maintain. The nominal orbit is defined by the ground. The orbit of a satellite is controlled by applying forces to the satellite. Forces are applied by delta-V *actuators* (so called because they impart a velocity change, or delta-V, to the satellite). Orbit control has traditionally been done in open loop mode under ground control. This means that the ground:

- determines the actual satellite orbit (by tracking the satellite as it passes over the ground station);
- computes the corrective forces that need to be applied to the satellite to bring its actual orbit closer to the nominal orbit;
- uplinks telecommands to the satellite commanding the application of the corrective forces to the orbit actuators.

Recently, there has been a trend towards moving the orbit control function to the satellite. In this case, an orbit control cycle would be defined similar to the attitude control cycle. Its frequency would, however, be much lower with typical periods ranging from several minutes to several hours.

Telecommand Processing Function

Telecommand packets are forwarded by the central satellite computer to the AOCS subsystem. telecommands arrive asynchronously. Upon reception by the satellite, some checks are performed on the telecommand (consistency with current operational status, verification of checksum, etc). If these are passed, the tele-command is then executed. Typical commands that are sent to the AOCS through telecommands include:

- Set nominal attitude;
- Command orbit control manoeuvre;
- Power up/down an AOCS unit (sensor or actuator);
- Command reconfiguration of AOCS units (e.g. fallback from prime to redun-dant unit, see section);
- Mark an AOCS unit as 'unhealthy' (not to be used in future reconfigurations);
- Command change of AOCS operational mode.

Some of the above commands override functions that can also be performed autonomously by the AOCS.

Telemetry Processing Function

AOCS cyclically generates telemetry to be sent to the satellite central computer. Depending on the system architecture, the forwarding of telemetry packets may be initiated by the AOCS itself or the packets may be autonomously acquired by the central computer. Telemetry cycles have low frequencies of the order of a fraction of an Hertz. The format of the telemetry packets (ie, the type of information that they contain) is determined by telecommands. Typical data that are contained in a telemetry packet include:

- power status of AOCS units;
- health status of AOCS units;
- current AOCS operational mode;
- telecommand log (ID of received telecommands, execution status, etc.);
- latest estimate of satellite attitude;
- latest value of control torques computed by the AOCS;
- latest readings from AOCS sensors;
- latest set of commands sent to the AOCS actuators;
- event log (log of 'special' occurrences such as reconfigurations)

Failure Detection and Isolation Function

This is a cyclical function where the AOCS attempts to detect anomalies and iso-late their cause. This function is very mission specific and can be very complex. Typical checks that are performed in order to detect anomalies include:

- *Attitude Anomaly Detection*: this check verifies that the attitude is within a pre-specified permissible window. The window may be defined with respect to the sun (e.g. "spacecraft X-axis shall not deviate by more that 20 degrees from the sun line") or with respect to the nominal attitude (e.g. "actual attitude shall not deviate from the nominal attitude by more than 25 degrees"). In both cases, a violation of the attitude anomaly check indicates that the attitude control function has failed. The size of the attitude window may be either fixed or settable by telecommand. The attitude anomaly check is a failure *detection* only method as it does not allow *isolation* of the cause of the anomaly.

- *Single Sensor Consistency Check*: this check verifies that the output of a sensor respects some physical constraints. Sensor outputs should for instance lie within certain ranges, and variations across two consecutive readings must be constrained to be below a certain threshold. The check thresholds are generally settable by telecommand. This check allows both detection and isolation of the failure.
- *Multi Sensor Consistency Check*: sensors usually provide redundant information (i.e. several sensors may be measuring the same physical quantity, for instance the sun direction). This check exploits the redundancy of the measurements to detect and, if the degree of redundancy is sufficient, isolate failures. The check thresholds are generally settable by telecommand.
- *Watchdog Alarm*: the AOCS software cyclically generates a 'watchdog event' that is detected by a dedicated hardware (the 'watchdog'). If the watchdog does not receive the event within a certain time-out, an anomaly is assumed to have occurred. This mechanism only allows fault detection, not fault isolation.

Failure Recovery Function

A failure recovery action is an action taken in response to the detection of a failure. Typical failure recovery actions include:

- *Failure Reporting*: The AOCS takes no autonomous action and simply reports the failure to the ground via telemetry.
- *Reconfiguration*: A reconfiguration is commanded (see next subsection). If the source of the failure has been isolated, the reconfiguration may be local to the equipment or function that has failed. If no failure isolation was possible, the entire subsystem may be reconfigured.
- *Mode Fall-Back*: A mode change is commanded to a lower-level operational mode in the expectation that the failure will not persist.

The failure recovery function may either be performed cyclically or be triggered in response to a failure detection.

Reconfiguration Function

At any given instant in time, the AOCS is using only a subset of all its available units. Note in particular that the redundancy requirement (section 4.5) dictates the presence of two sets of each sensor. Hence, not more than half of all available equipment are ever used[4]. When the AOCS has detected a failure, it must attempt

---

[4] This is strictly speaking not true because there are units, like the reaction wheels or the gyros, where redundancy is achieved without full duplication of hardware. Thus, for instance, in a set of four gyros oriented in such a way that no three gyros have coplanar

to recover by performing a reconfiguration. A reconfiguration changes the set of units that are being used by the AOCS at a certain instant in time. The intention of a reconfiguration is to exclude the faulty unit. The reconfiguration logic is highly mission dependent. However two basic types of reconfigurations can be identified:

- *Unit Reconfiguration*: If the failure detection and isolation function was able to isolate the faulty unit, then the typical reconfiguration is a switch-over to the redundant copy of the faulty unit. If, for instance, a consistency check on the output of a sun sensor shows that the sensor is defective, then this function will switch over to the redundant sun sensor. Overall system performance should remain unaffected.
- *Subsystem Reconfiguration*: If no fault isolation was possible, the entire subsystem is reconfigured. This means that all units (including the AOCS computer) are switched over to their redundant copies and the subsystem is re-initialized. If, for instance, the attitude anomaly safeguard has been violated, then it is known that a serious fault has occurred but it is not possible to say which equipment is responsible for it. In this case, all currently used hardware is switched off and there is a switch over to the redundant sets.

Manoeuvre Management Function

A manoeuvre is a sequence of actions that must be performed by the AOCS at specified times to achieve a specified goal. Examples of manoeuvres include:

- A sequence of small torque impulses imparted to the spacecraft to cause the speed of the reaction wheels to change in a desired manner (*wheel momentum unloading*).
- A slew to change the direction of pointing of the spacecraft payload.
- A sequence of thruster firings to change the spacecraft orbit in open-loop mode.

Manoeuvres can either be triggered autonomously by the AOCS or they can be commanded by the ground via telecommands. Manoeuvre execution may overlap, with more than one manoeuvre being active at the same time.

2.1.2 AOCS Operational Modes

The great variety of external conditions under which a satellite must operate in a given mission, dictates the breakdown of the overall AOCS functionality into several operational modes with each operational mode optimized for certain mission conditions. The functionalities of the AOCS are always the same as explained in the previous section but their *implementation* changes from operational mode to operational mode. Consider for instance the attitude control function. The accuracy with which the satellite attitude must be aligned to the nominal attitude varies

---

sensing axes, there is 3-out-4 redundancy in the sense that any three gyros provide full angular rate information.

form mission phase to mission phase (and hence from operational mode to operational mode). Consequently, different control laws are used to implement this function depending on the required accuracy.

An important difference between operational modes lies in the set of AOCS units – the sensors and actuators – that are used by the AOCS. In principle, each operational mode uses a different set of units. Hence, a mode transition is usually accompanied by a power-up and power-down sequence for some units.

Mode transitions can either be commanded by the ground by telecommand or they can be decided autonomously by the AOCS software. In the latter case, however, the ground retains the right to over-ride or inhibit AOCS-initiated transitions. The mode architectures is very mission specific. However, on most missions, some or all of the modes listed in table 3 would be found. Each of the modes in the table may in turn be divided into sub-modes. The safe mode in particular will usually contain its own version of the rate reduction and attitude acquisition.

Table 3. Typical operational modes of an AOCS

AOCS Mode	Mode Description
Stand-By Mode	Only telecommand and telemetry functions are active. No attitude or orbit control is performed. All the AOCS units, except the AOCS computer, are switched off. Mode is typically used when satellite is still attached to the launcher or in the first seconds after separation from launcher.
Rate Reduction Mode	Objective is to reduce angular velocity to a very small value (of the order of 0.01 deg/sec). More is typically used after separation from the launcher (launcher separation induces comparatively high angular rates on the satellite).
Attitude Acquisition	Satellite is slewed until it is oriented according to a pre-defined attitude. Target attitude is often selected to have solar arrays sun-pointing (to ensure steady power supply) and radiofrequency antennas earth-pointed (to ensure good telecommand and telemetry link).
Fine Pointing Mode	High accuracy mode where satellite attitude is finely controlled. Mode is typically used during payload operation.
Slewing Mode	Mode used to change satellite orientation (for instance, when a telescope must be pointed to a new target).
Orbit Control Mode	Mode used when forces are imparted to the satellite to change its orbit.
Safe Mode	Emergency mode entered when a failure is detected and no reconfiguration is possible to counter the failure. Objective is to keep satellite in an attitude in which no permanent damage to the satellite or its instruments can occur. Payload operation is suspended.

2.1.3 AOCS Units

The AOCS units are the AOCS sensors and the AOCS actuators. The *sensors* are used by the AOCS to collect measurements about the current attitude and position of the satellite. The *actuators* are used to impart torques and forces to the satellite to control, respectively, the satellite's attitude and its position. Attitude sensors can be of two types. *Passive sensors* have no internal processor. *Active sensors* have an internal processor and the complexity of their software can match that of the AOCS computer itself. In some architectures (proposed but so far not implemented), some of this software is located on the AOCS computer. The most common types of passive sensors are:

– *Fine Sun Sensor (FSS)*: A fine sun sensor measures the direction of the sun line in the sensor's reference frame Typically, FSS's have a nearly-hemispherical field of view and very high accuracy (order of arcsecond).
– *Coarse Sun Sensor(CSS)*: The CSS is conceptually similar to the FSS but it is constructed in a different way. The different technology results in a lower accuracy (order of 0.1 deg). The field of view is nearly hemispherical with accuracy being highest close to the boresight.
– *Sun Presence Sensor (SPS)*: This sensor has a binary output. Possible outputs are: "sun present" and "sun not present". The "sun present" output is generated when the sun direction is within the sensor's field of view. Typically the field of view is square with a side of 25 or 30 degrees. The "sun not present" output is generated when the sun is outside the field of view. SPS's are often used as attitude anomaly detectors.
– *Earth Sensor (ES)*: Earth sensors measures the earth direction in the sensor's field of view. Their accuracy is rather low (order of 0.01 to 0.1 deg).
– *Magnetometers (MGM)*: Magnetometers measure the direction of the earth's magnetic field in the sensor's field of view. Their accuracy is low and varies widely from system to system.
– *Gyroscope (GYR)*: Gyroscopes measure the inertial angular rate of the satellite. Gyroscopes can have one or two sensitive axes. They give the projection of the satellite angular rate on the sensitive axes. Gyroscopes are often combined in packages to give measurements along up to six axes.

Active attitude sensors generate measurements that are obtained by performing sophisticated processing on the raw measurements generated by the sensor's hardware. In the conventional architecture, the sensor's software runs entirely on the sensor's processor. Some proposed architectures instead envisage overcoming processing limitations on the sensor by placing part of its software on the AOCS computer. Obviously, in such cases, the distinction between sensor software and AOCS software can become blurred. It would exist mainly from a project management point of view as the sensor software is supplied by the sensor supplier who is in general different from the supplier of the AOCS software. There are at present two types of active AOCS sensors:

- *Star Sensor (STR)*: A star sensor measures the positions of several stars in its field of view. It also performs pattern recognition on the stars it sees to identify the portion of sky at which it is looking.
- *GPS Receiver*: GPS receivers are primarily used for position determination but they can also provide an inertial measurement of the host satellite attitude.

Position sensors measure the position of the satellite. Position measurements are required as input to the orbit control function. There is at present only one position sensor. On low earth orbits, a GPS receiver can provide an estimate of the host satellite position to an accuracy of 50 m or better. Indirect position measurements are also possible by combining measurements from star, earth and sun sensors.

If we now consider the actuators, we find that there are three basic types of AOCS actuators:

- *Thrusters (THU)*: Thrusters can emit gas jets that impart a force to the satellite. If the direction of the gas jet does not go through the satellite centre of gravity, then a torque on the satellite results. By combining jets from suitably located thrusters it is possible to apply either pure forces or pure torques to the satellite. Hence thrusters can serve either as attitude or position actuators.
- *Reaction Wheels (RW)*: A reaction wheel is a rotating wheel that can be accelerated or braked. The action of accelerating or braking causes a torque to be applied to the satellite by reaction. Conceptually, reaction wheels are devices that can apply a torque along one axis.
- *Magnetorquers (MGT)*: Magnetorquers are devices that interact with the earth's magnetic field to generate a control torque.

2.1.4 The AOCS Software

The AOCS software is generally built as a single load module that is burned into PROM on the AOCS computer or uplinked by telecommand. For power reasons, the software normally runs from RAM and patches to the software are therefore possible after launch. The size and complexity of the AOCS software has in the past been limited by the available hardware. Processors for use in space must undergo a lengthy and expensive qualification process that certifies their robustness to the space environment (radiation, launch shock, temperature range, etc.). The traditional processor used in ESA space missions implements the mil-std 1750 architecture. Available memory in the "standard" configuration is only 64 Kbytes and many AOCS load modules had to fit within this rather constrained space. Recently, a more advanced processor, the ERC32, has been qualified for use in space that implements the SPARC architecture and can have several Mbytes of memory [37]. Similar developments have occurred in the US space industry.

The language of choice for the AOCS software on ESA projects is ADA83. The AOCS software is organized as a collection of tasks. A cyclic scheduling policy without pre-emption is used. The cycle period is the same as the AOCS control cycle (often 500 ms but also as much as 1 sec or as little as 50 ms). Tasks are allocated a slot in this cycle and must return within the allotted time. Overruns are normally fatal errors causing a reset of the software. Ada tasking is generally not used because of its overheads. Interfaces with the hardware are interrupt-based.

The AOCS software has two major hardware interfaces: with the subsystem bus and with the system bus. Through the former, data exchanges with the AOCS units are routed. Through the latter interface, telemetry and telecommands are collected and received. On the subsystem bus, the AOCS computer is the master and initiates all communications. Data exchanges with the subsystem bus can be either through I/O instructions or through DMA. When the AOCS computer initiates a data exchange on the subsystem bus, interrupts may be used to notify the AOCS software of the termination of the data exchange or of the arrival of a response from a unit. On the system bus, the AOCS computer normally behaves as a slave. The arrival of telecommands can be signalled by an interrupt. Telemetry is instead stored by the AOCS computer in a dedicated buffer from which it can then be autonomously collected by the system bus interface in DMA mode. Finally, the occurrence of exceptions (e.g. divide-by-zero errors, hardware errors, etc) is usually not handled and simply causes a complete reset of the software.

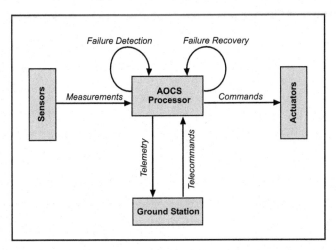

Fig. 2.3. Conceptual structure of an attitude and orbit control system

2.2 The AOCS and Other Control Systems

Figure 2.3 shows the conceptual architecture of an AOCS. The horizontal data flow is the most important one. It represents the processing of sensor measurements to generate actuator commands and the generation of these commands is the ultimate purpose of the AOCS. The vertical data flow represents the interaction with the ground station. Failure detection and failure recovery are instead shown as actions that the software performs upon itself. Except for the commands from the ground station, all the data flow and the software actions have a periodic character that reflects the cyclical nature of the AOCS. The structure shown in the figure is that typical of an embedded control system with the only minor difference that the ground station is replaced by an external operator or by some supervisory computer. It is this similarity of the AOCS to a generic embedded control system that justifies our claim that many of the constructs offered by the AOCS Framework – in particular many of its design patterns – could be reused in other embedded control contexts.

3 Software Frameworks

In introducing software frameworks, it is useful to distinguish between their purpose – what they are for – and their definition – what they are. Broadly speaking, software frameworks are a form of software reuse that primarily promotes the reuse of entire architectures within a narrowly defined application domain. They propose an architecture that is optimized for all applications within this domain and make its reuse across applications in the domain possible. Experienced software engineers who develop several related applications reuse architectures as a matter of course. Software frameworks are intended to formalize and make explicit this form of architectural reuse. They allow the investment that is made into designing the architecture of an application to be made available across projects and across design teams.

If there is a general consensus over the purposes of framework technology as briefly summarized above, disagreement instead reigns over what concretely constitutes a framework. Roughly speaking, the conceptualizations that can be found in the technical literature can be arranged on a continuum with, at one end, an architectural-centered view of software frameworks and, at the other end, a component-centered view. Researchers who consider frameworks from a theoretical point of view tend to be at the architectural-centered end of this spectrum. An example is [41] that describes object-oriented frameworks as "collections of components with predefined collaborations between them and extension interfaces" (pag. 10). [25] similarly sees a framework[5] as a set of components capable of being assembled into an application under the umbrella of a common architecture which is regarded as the "key core asset" of the framework.

The view of frameworks that is behind such definitions is probably that shown in figure 3.1. The unshaded area represents the architectural backbone shared by all applications in the framework domain. The framework captures this architectural backbone and makes it available to application developers who tailor it to their needs by plugging into the framework components that implement the application-specific behaviours (the darker boxes in the figure). This view emphasizes the character of frameworks as customizable constructs. As suggested by the figure, one of their peculiarities then lies in the fact that the application-specific code is called by the reused code. This is made explicit in [41] where frameworks are defined by their reliance on the "call-back style of programming" as the primary

---

[5] This and other references in this section are concerned with product lines. There is as usual much terminological confusion but in this report it will be assumed that product lines are built around a framework.

A. Pasetti: Software Frameworks, LNCS 2231, pp. 29-47, 2002.

functionality extension mechanism. Inversion of control is seen as a key feature of frameworks in [39] and [51], too, which offer conceptualizations of frameworks similar to that of [41] as collaborative models of component interactions together with suites of components to allow code as well as architectural reuse. Adaptation through call-back is in contrast to older reuse techniques, like for instance subroutines or class libraries, where instead the application-specific code was in control of execution and was responsible for calling the reused code.

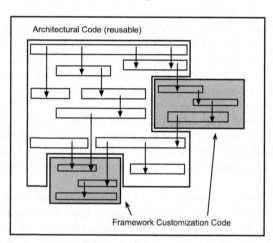

Fig. 3.1. A first view of software frameworks

Figure 3.1 also represents well the view advocated in [4] where frameworks are defined as "variance-free reference architectures" plus a large set of pre-defined plug-in components. References [33, 93, 82] follow the same trend in explicitly differentiating component repositories from frameworks with the latter's emphasis on providing an architecture as well as a set of reusable components. All these authors probably agree that an important distinctive feature of software frameworks is that they do not simply supply individual components (or individual classes) but also provide the *interconnections* among the components they supply. It is this web of interconnections that constitutes, in an abstract sense, the architecture whose reuse is made possible by the framework.

If one moves towards the component-centered end of the framework conceptualization continuum, one finds authors who identify the framework with a set of closely related components. This is, for instance, the case of [99] who sees a framework as a "topology" of components. The emphasis however it now more on the components than on any underlying architecture since code is regarded as a valuable (the most valuable?) asset. The SPLIT method of [30] takes a similar view in regarding a framework as made up of components together with an architecture to ensure conceptual integrity. Reference [16] also models frameworks as families of components within a reference architecture. Reference [97] moves further towards the component end of the framework continuum by formally defining a framework as a set of abstract and concrete classes.

Experience reports on the use of frameworks in concrete projects tend to be at the component-centered end of the framework continuum. Examples can be found in [63, 31, 67]. Often, these authors more or less tacitly operationalize a framework as little more than a component library that facilitates the development of applications within a certain domain.

The framework conceptualization proposed in this book is definitely at the architecture-centered end of the framework continuum. It thus shares with the previous definitions the view that a software framework is primarily a means of reusing an entire architecture. These definitions however suffer from one drawback in that they remain somewhat vague about what exactly constitutes an architecture. Reference [41] for instance does not make explicit how the predefined collaborations and extension interfaces that are seen as the distinctive feature of a framework with respect to a mere component library are realized in practice. They obviously represent the "architecture" that the framework captures and makes available for reuse but the vehicle through which they are imported into target applications is not stated. Reference [25] lists "architectural patterns and styles" among the inputs that can shape the framework core assets but it does not elaborate the point and their role is not clarified. This could range from the development of domain-specific patterns, to the adoption of catalogue patterns, or even to their use simply as a means of documenting the framework design and of helping to communicate its structure. References [39, 53] come closest to the view presented below in seeing design patterns together with components as the architectural elements of frameworks although it is never clear whether they refer to generic patterns or to patterns that are domain-specific and whose creation then becomes one of the chief contributions of a framework. Similarly, reference [93] stresses the role of design patterns in building architectures but sees them as "encapsulation of best practice" and the specific patterns that are proposed to organize the framework are not domain-specific.

As will be seen, the concept of framework that is advocated here is characterized by an emphasis on domain-specific design patterns in framework design and use. The patterns are seen as the basis upon which the framework is constructed, as the vehicle through which architecture is made reusable, and as the means to broaden the applicability of the framework approach to domains where internal complexity is such that it is not possible to define a single rigid reference architecture.

Before proceeding to discussing this concept of framework, it is perhaps useful to briefly present an alternative view of frameworks as shown in figure 3.2. The figure distinguishes two levels – the *framework level* and the *application level* – to indicate that frameworks and applications are conceptually different and exist at different levels of abstraction. It shows that a framework models a whole set of related applications and suggests that frameworks can be seen as *generative devices*, namely as starting points for the construction of individual applications in the framework domain. The process whereby a concrete application is derived from the framework is often called *framework instantiation*. This is the process that governs the transition from the upper to the lower plane in the figure. This second view of frameworks is related to the role proposed for them in the application development environment presented in the first chapter (see figure 1.3) where they are to complement tools likes Xmath/Matlab to automatize the instantiation of applications within the framework domain.

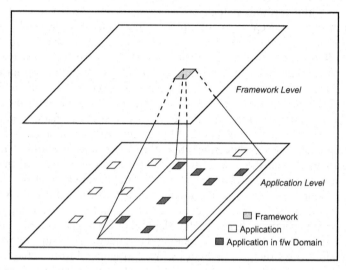

Fig. 3.2. A view of frameworks as generative devices for applications

The work done by D. Batory – one of the framework pioneers – is perhaps the one the follows most closely the conceptualization of frameworks of figure 3.2. It sees frameworks as collections of components and it treats code reuse as very important. Reuse, however, occurs within the confines of a kind of domain grammar that formally specifies how the components can be composed both from a syntactic and semantic point of view. Composition can moreover, at least in principle, be automatized thus resulting in the very rapid and very safe instantiation of individual applications. To the extent that the domain grammar implicitly defines an architecture, this approach achieves a very high degree of both architectural and code reuse. The price it pays for this remarkable accomplishment is narrowness of applicability since formal domain grammars can be defined only for very structured domains whose applications can be explained entirely in terms of composition of primitive, plug-compatible and interchangeable components. The example framework mentioned in [13], for instance, targets graph-traversal applications. The application in the ADAGE project and reported in [10, 11], although nominally aimed at avionics systems, is in reality restricted to the data flow from sensors to actuators. It is clear that domains like that of AOCS applications considered in the case study in the second part of the book are far more complex and less susceptible to formalization. It is possible, though, that even this domain could be decomposed into subdomains to which ADAGE-like approaches could be individually applied. This possibility, however, is not explored in this book.

3.1 Frameworks, Components, and Inheritance

Frameworks are artifacts that are intended to be customized for use in a specific application. Following a widespread convention introduced in [76] the points where the customization takes place are called *hot-spots*. The type of mechanism through which the adaptation takes place (object composition, inheritance, parameter tuning, instantiation of generic packages, etc) serves to distinguish among categories of frameworks.

Consider again figure 3.1. The concept that the figure wants to convey is that of a framework as a collection of entities together with their connections. The nature of these entities depends on the particular technology that is used to implement the framework and frameworks are compatible with many different technologies. Thus, the small rectangles in the figure may represent subroutines, modules, objects, classes, etc, depending on the framework implementation technology. This book is concerned with *component-based frameworks*, namely frameworks whose constitutive elements are *components*. The term "component" is yet another of those heavily over-loaded terms that need to be carefully defined before use. In this book, it will designate an entity that possesses the following properties:

- it can be compiled as a stand-alone unit
- it exposes one or more interfaces
- it provides an implementation for the interfaces it exposes
- it is seen by its clients only through the interfaces it exposes

The idea in the first bullet is often expressed by saying that a component is a "binary unit of reuse" [98]. The natural customization mechanism for a component-based framework is object-composition. Thus, if one wants to continue with the metaphor of figure 3.1, its hot-spots should be imagined as holes whose shape is defined by an interface. A customization component can be inserted into a certain hole only if it has a matching shape or, in other words, if it implements the right interface.

Frameworks can and have been built for purely object-based systems. A prominent example in the space field is the OBOSS framework [71] that is targeted at Ada83 applications. Most frameworks under development at present, however, tend to rely, at least in part, on inheritance as a behaviour adaptation mechanism. In the case of the framework for embedded control systems discussed later in the book, inheritance is often from abstract base classes.

In general, a framework can be inheritance- and component-based at the same time[6]. This simply means that behaviour adaptation is achieved partly through object composition and partly through inheritance. The two adaptation mechanisms are to some extent interchangeable. It has been remarked [3] that frame-

---

[6] The terms "white-box" and "black-box" are often used for inheritance-based and component-based frameworks. They refer to the fact that use of inheritance requires more detailed knowledge of the internal structure of the framework (i.e. it requires the framework to be "white" or "transparent") than is the case when using object composition.

works tend to be initially based on inheritance and then evolve towards greater use of object composition. This trend can be explained with the fact that inheritance is more convenient from the point of view of the framework designer but harder to use for the application developer (the user of the framework) whereas the opposite applies to object composition. The framework for embedded control systems presented in this book has a mixed component-based and object-oriented character.

With reference to figure 3.2, one can imagine that the "framework level" is the plane where the abstract interfaces implemented by the components and abstract base classes are defined and the "application level" is the plane where the concrete components and classes are defined. The process of instantiation then becomes the process of instantiating the framework's abstract classes into concrete classes and implementing the abstract interfaces in components. If one considers the design patterns introduced in the second part of the book, it will be noticed that they often exhibit a two-level structure with framework classes and interfaces at the top and concrete application classes at the bottom. This structure mirrors the structure shown in the figure and is regarded as typical of object-oriented frameworks that use inheritance and object composition as adaptation mechanisms.

3.2 A More Complete View of Software Frameworks

One way to arrive at a definition of a software framework is to ask the following question: if one were to buy a software framework, what would one get, in practical terms, in return for one's money? Which products or services would be specified in the purchase contract? It is noteworthy that many of the definitions that are provided in the technical literature do not allow to give precise answers to these questions mainly because they fail to define precisely the concept of architecture which, in most cases, they insist is an important part of the framework. The answer that is proposed here is that a framework should be constituted of three types of constructs:

– abstract interfaces
– concrete components
– domain-specific design patterns

A framework as the term is understood in this book is then a set of related abstract interfaces, design patterns and concrete components that are designed to be used with each other to facilitate the development of applications in a particular domain. These constructs can therefore be seen as the set of tools that the framework makes available to application developers to assist them in building a particular application in the framework domain. The case study in the second part of the book shows how these three categories of constructs are composed to capture the invariant part of the architecture of satellite control systems.

The *abstract interfaces* consist of declarations of set of related operations for which no implementation is provided. Abstract interfaces capture behavioural signatures that are common to all applications in the framework domain. In a language like C++, an interface is represented by a class with pure virtual methods only. Other languages – typically Java – instead explicitly define the concept of interface as a primitive language concept.

Abstract interfaces play two roles in the framework. On the one hand, they can represent framework hot-spots through which the framework is adapted by inheritance. On the other hand, they are used to create links between components within the framework through abstract coupling. Abstract coupling is heavily used in object-oriented frameworks to create clean separations between interacting components and to achieve extensibility.

Note that, in principle, abstract interfaces could be subsumed under the category of design patterns since an abstract interface can be regarded as a small and trivial design pattern. This was not done to avoid the confusion that would inevitably arise from a redefinition of such expressions as "design pattern" and "abstract interface". However, readers should bear their conceptual similarity in mind and the discussions later in this book about the role of design patterns in the development of frameworks will often implicitly include abstract interfaces under the same umbrella.

The *concrete components* – the second of the three basic framework constructs – are components that are provided by the framework. The framework components can be of two types: *core components* and *default components*. The *core components* encapsulate behaviours that are common to all applications in the framework domain. Thus, the core components are used by all applications instantiated from the framework. *Default components* represent default implementations for some of the abstract interfaces in the framework. They encapsulate behaviours that are found in many applications in the framework domain but that are not intrinsic to the framework domain. A framework will in general provide default components to implement the most common behaviours found in its domain. When very specific behaviours are required that are not catered for by the default components, application-specific components need to be created. These will have to conform to the framework interfaces and may often be obtained by specializing the default components either through inheritance or through composition.

The *design patterns* are the third basic construct provided by a software framework and, despite being listed as last, they are conceptually the most important. In general, a *pattern* is a perspective on a subject expressing a general recurrent theme that has proven to be useful. More specifically, a *design* pattern has software design as its subject and represents an optimized solution to a recurring software problem [51]. The expression is used in the same and well-known sense of [44, 24]. Software engineers are nowadays familiar with the concept of design patterns but are normally acquainted only with general purpose patterns. A framework as understood here instead contributes domain-specific design patterns, namely design patterns that either address problems that are specific to its domain or that address common problems but propose solutions that are optimized for the framework domain.

Good design patterns are both valuable and hard to find. The view advocated here is that, just as software engineers have been in the habit of maintaining libraries of subroutines and components, they should learn to do the same for design patterns. Frameworks are one vehicles through which such repositories of design patterns can be built and made available to several users.

In the conceptualization proposed here, design patterns lie at the very heart of frameworks. Frameworks are intended to make architecture reusable and design patterns encapsulate atomic architectural solutions. Additionally, design patterns promote architectural uniformity by ensuring that similar problems in different parts of the same application receive similar solutions. They endow the application architecture with a single "look & feel" that makes using and expanding it considerably easier. Design uniformity is an important aspect of architectural reusability and a further contribution of design patterns to frameworks. Standard catalogues of patterns offer design solutions that are generic and which are seldom "ready to use" (they normally leave several implementation and design options open to the user). Frameworks should instead offer design patterns that, though possibly inspired by standard patterns, have been refined and specialized for use in a specific domain. The framework patterns should be as narrowly defined as the variability within the domain allows. One of their functions is to save application developers the trouble of inspecting standard pattern catalogues to identify relevant standard patterns and, where necessary, to adapt them to their needs.

At this point, it is useful to introduce the notion of *instantiated design pattern*. A design pattern proposes a *design* solution. After being instantiated, the design pattern represents an *architectural solution*. An example will clarify this distinction. Considers for instance the well-known Factory Method pattern. Its standard UML diagram as found in [44] is:

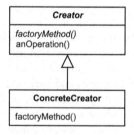

Pattern instantiation takes place when the pattern is applied to a concrete architectural situation and results in the two classes of the pattern being mapped to specific classes in a specific project and in other non-essential changes (like for instance having more than one factory methods).

Design patterns can enter a framework in two guises: as *instantiated patterns* and as *exportable patterns*. Design patterns that are instantiated in the framework are design patterns that are used within the framework itself. Exportable design patterns, by contrast, are design patterns that are offered to the application developer for him to instantiate them in his architecture. Note that the same design pattern may be both instantiated within the framework and offered for instantiation to application developers. One example from the AOCS Framework is the Mode

Management pattern (see chapter 12) that is extensively used within the framework to provide selected components with mode-dependent behaviour and that is at the same time exported to application developers for them to build it into their own mode-dependent components.

It is a commonplace to say that frameworks promote architectural reuse but few authors specify how exactly architecture is to be captured and made available for reuse. The answer provided here is that architectural solutions are encapsulated in design patterns which are either used within the framework (instantiated design patterns) or exported to application developers to help them construct their applications (exported design patterns).

Finally, it should be noted that the way the expression *design pattern* is used in this book encompasses concepts like architectural patterns or architectural style as defined in texts like [92, 20]. The latter author in particular reserves the term *design pattern* only for localized design solutions. Here instead it is used more broadly to apply to all context-independent design solutions irrespective of their granularity. This issue is considered again below.

The three sets of constructs listed at the beginning of this section are intended to constitute a minimal description of a framework in the sense that their definition should be both necessary and sufficient to fully describe the framework. Note in particular that the conceptualization of software frameworks proposed here does not include hot-spots among their primitive constructs. This is because hot-spots are intrinsic to either abstract interfaces or design patterns. Once the abstract interfaces and the design patterns of the framework have been defined, the hot-spots are also implicitly defined.

The view of software frameworks proposed here shares with those discussed at the beginning of the chapter the emphasis on architectural reuse as the essence of the framework approach. However, in promoting design patterns to the rank of primary framework constructs, it is more explicit about how architectural information is to be represented within the framework and how it is to be reused by application developers. It is also more flexible because the conceptualization of a framework as a set of domain-specific design patterns allows modelling a larger variety of systems. The model of figure 3.1 is suggestive and simple to use but remains restricted to very narrow domains where all applications share the same architectural skeleton and where application instantiation is reducible to filling in hot-spots with appropriate components. This is possible only for simple and fixed-structure domains. In more complex domains, application developers will have to do more than just plug-and-play. Our conceptualization is suitable for such situations where the target domain is characterized not so much by a single reference architecture but more broadly by a set of shared design patterns. To some extent, the design patterns become building blocks *for* an architecture. Other approaches, instead, rely on components being used as building blocks *within* a rigid architecture. The reuse they promote, in other words, tends to be at code and architectural level only whereas our conceptualization explicitly allows and fosters reuse at the design level. This has the advantage of making it applicable to complex domains like that of satellite control systems explored in the case study. Obviously, this advantage is partially offset by the greater burden it places on the application developers who will not be able to simply assemble their application by combining lego-like blocks.

The issue of the levels of reuse allowed by a framework deserves a closer look. Figure 3.3 shows the levels of reuse that are relevant in a framework context. At the highest level of abstraction, only the design is re-used by which it is meant that the elements of reuse are the design patterns. At the next lower level, one can reuse an architecture (either in its entirety or in part). In that case, the focus of reuse are the instantiated design patterns that encapsulate the architecture. In practice, the reused elements are the definitions of the component interfaces. Finally, at the lowest level of reuse there is code reuse that consists in directly reusing concrete components or concrete classes. Traditional reuse technologies have been slanted towards code reuse. More recently, architecture reuse has been proposed. Design reuse has been practiced implicitly but has seldom been recognized as a distinct level of reuse. Here it is proposed that a software framework should span all three levels of reuse and that only by combining reuse at all three levels and by focusing on a narrow domain can reuse really pay off.

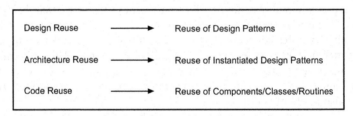

Fig. 3.3. The three levels of software reuse

Other authors have also recognized the multi-layered character of software reuse. Reference [39] remarks that design patterns allow pure design reuse and that frameworks are intermediate between them and component/subroutine libraries in offering a mixture of design and code reuse. Similarly, reference [51] identifies three levels of reuse: *implementation,* which is essentially identical to code reuse; *design,* which conflates the architectural and design reuse as understood here; and *domain,* that concerns the reuse of domain models. Our reuse model is characterized by an explicit distinction between design and architectural reuse. This distinction is important for, as discussed below, it reflected in a distinction between architecture and design as distinct phases in the development of a framework.

It should be noted that in domains where it is possible to define a single reference architecture, design-level reuse becomes identical to architectural-level reuse. It is probably because most research and practice in software frameworks has tended to assume the existence of such a reference architecture that the distinction between the two levels of reuse is often blurred. In our work on embedded control systems it came to the forefront because the complexity of this domain simply does not allow the definition of one (or even a small number of) reference architectures. On the other hand, anyone familiar with control system applications will recognize that certain patterns tend to recur and where there is a pattern, there is the potential for reuse. This potential cannot be tapped by the framework approaches discussed at the beginning of the chapter because it cannot be entirely reduced to component or even architectural reuse. Individual components and architectural fragments are indeed reusable but in addition it is possible to have reuse at the design level through the reuse of design patterns. One of the chief contributions of our conceptualization is to extend the framework approach to this type of domains by expressly making reuse possible at code, architectural and design level.

Finally, this chapter has introduced many terms to designate various aspects of a framework-based software development process. For ease of reference, they are presented in table 4 together with their definition as adopted here. As already noted many of these terms are overloaded and other authors would use them with different meanings.

Table 4. Definition of key framework terms

Term	Definition
Architecture Reuse	Reuse of the definitions of component interfaces or, equivalently, reuse of instantiated design patterns.
Code Reuse	Reuse of concrete components.
Component	A software module that can be compiled as a stand-alone unit and is characterized by one or more interfaces that it exposes to its clients and for which it provides implementations.
Core Component	A component provided by a framework that encapsulate behaviour that is common to all applications in the framework domain.
Default Component	A component provided by a framework representing default implementations for some of the abstract interfaces in the framework
Design Pattern	An optimized solution for a recurring design problem.
Design Reuse	Reuse of design patterns.
Domain-Specific D.P.	A design pattern that is optimized for use within a specific domain.
Exportable D.P.	A design pattern that a framework offers to application developers for them to instantiate in their applications.
Framework	A set of related abstract interfaces, domain-specific design patterns and concrete components that are designed to be used with each other to facilitate the development of applications in a particular domain.
Framework Domain	The set of applications that the framework can help generate.
Framework Instantiation	The process whereby a framework is transformed into a concrete application within its domain.
Hot-Spot	A point where the framework can be adapted to match the needs of an application to be instantiated from it.
Instantiated D.P.	A design pattern that is used within the framework itself.

D.P. = Design Pattern

3.3 Frameworks and Autocoding Tools

The analogy between frameworks and autocoding tools, at least in the embedded control domain, was already noted in chapter 1. Autocoding tools like Xmath or Matlab allow control engineers to design a control system in a graphical environment using high-level concepts and to generate code that implements the design. There is a similarity with software frameworks because both technologies aim at reusing design even if the ways in which reuse is achieved are different. The conceptual differences between the two approaches deserve to be stressed because some see the two technologies are competing alternatives when in fact, as argued below, they are complementary[7]. There are at least five respects in which significant differences can be identified. The discussion below uses examples drawn from the AOCS Framework case study but its conclusions are quite general.

Firstly, an autocode tool does not by itself facilitate the design of the software architecture. *Given the architecture*, the tool makes it easy to generate the software but the design of the architecture has to be done manually by the user on an *ad hoc* basis. From a software point of view, the autocoding tool can be seen as a repository of domain-specific components. The view of frameworks advocated above instead sees them as components embedded within a pre-defined architecture. Use of a framework therefore relieves developers of the need to define the architecture of their software because the architecture is built into the framework (indeed, the framework *is* the architecture). This is a significant advantage because the architectural design is the hardest part in the development process and one whose complexity will grow as the complexity of the underlying system grows.

A second difference stems from the greater generality of the autocoding tools that are targeted at a very wide range of users (e.g. Xmath and Matlab are targeted at dynamic system modeling in general). Frameworks are instead targeted at specific application domains (e.g. the AOCS Framework is targeted at satellite control systems) and would therefore be endowed with abstractions that are tailored to this application domain. The availability of higher-level abstractions makes the user's code simpler and more readable, and hence more re-usable and easier to maintain. Consider for instance the AOCS domain. A generic autocode tool will not have an abstraction for 'sun sensor' because such a component is not required in the vast majority of dynamical system applications although it plays a prominent role in AOCS applications. The component must therefore be built from simpler entities for each new AOCS application. An AOCS Framework would instead be likely to include a 'sun sensor' abstraction that can be directly manipulated by users resulting in greater ease of use and in more readable code. The generality of autocoding tools also implies that they seldom can generate more than a fraction of the total code required for the AOCS. An AOCS-specific framework would instead cover the majority of the AOCS code.

---

[7] Some of the reviewers of the AOCS Framework, for instance, argued that development of a software framework for an embedded control system was unnecessary because tools like Matlab and Xmath already made architectural and design reuse possible and easy.

In a third difference, autocoding tools generate code that is hard to understand and with an obscure structure largely outside the control of the user. This becomes a problem when debugging is required or if code patches are applied during normal operations. Obviously, this problem is alleviated by a framework approach.

Fourthly, autocoding tools are usually built on top of environments that were intended as *simulation* and *algorithm* design environments, not as design environments for the software architecture of complex systems. The facilities they provide are correspondingly optimized for the former tasks and make them awkward to use as architectural design tools. One important consequence is that code or model reuse across projects is very difficult: understanding a complex Xmath model can be as challenging as understanding a complex piece of code. Frameworks, on the other hand, are specifically designed to be portable across projects and to have a structure that facilitates maintainability, re-use and understandability. A closely related point is the lack of support of autocoding tools for complex data structures which often makes architectural design unnecessarily complex and error-prone.

Finally, the origin of autocoding environments as simulation tools is also their greatest strength. An autocoding system provides a seamless integration of validation and development facilities. Although a framework that is based on components could also be oriented towards integrating development and testing environment, it is clear that at present a framework solution is in this respect definitely inferior to an autocoding solution.

Despite, or perhaps because of, the above differences it is necessary to stress that the autocoding and framework approaches are hardly incompatible. Indeed, they can be seen as complementary. Autocoding tools are especially good at generating code that implements specific algorithms (i.e. they are good at generating individual procedures, rather than entire systems) and frameworks typically have 'holes' precisely for the procedures implementing the application-specific algorithms. An AOCS Framework, for instance, will have handles where users can hook the procedures implementing, say, the controller algorithms and the state estimators. These algorithms would typically be developed and tested in an environment like Xmath or Matlab. These environments could be used to generate the code implementing them and this code could then be inserted in the appropriate 'holes' in the framework. This solution would optimally combine the framework and autocoding approaches and would leverage the specific advantages of each. In recognition of the complementarity of the autocoding and framework approach, the AOCS Framework presented in this book includes an interface to Xmath-generated code. This type of interface is also seen as a precursor to the integration of the autocoding and framework approach advocated in chapter 1.

3.4 The Methodological Problem

One of the contributions of this book is to present a set of guidelines and conceptual tools to aid the design of a software framework. Before proceeding to presenting them, it is necessary to explain why new methodological tools and concepts specifically targeted at software frameworks are required at all. Why, in other words, can framework development not be based on one of the many established design methodologies? In attempting an answer to this question, the term "conventional software methodology" will be used to designate the software development methodologies that have been proposed in recent years and that are now in wide use in the software industry. Prominent examples of such methodologies include OMT [88], the approach originated by Booch in [17, 18], that proposed by Shlaer-Mellor [94], and the COAD-Yourdon [26] and HOOD methodologies [49]. There are several strong reasons why these methodologies are inadequate for the design of a software framework. Before expounding them, it must be stressed that their inadequacy is predicated upon the particular conceptualization of software framework adopted in this book. It is obviously possible that adoption of a weaker (probably *much* weaker) interpretation of the term "framework" might restore the compatibility between frameworks and conventional software methodologies.

3.4.1 Design Patterns and Abstract Interfaces Vs. Concrete Objects

Of the many ways in which a software artifact can be conceptualized, conventional methodologies tend to privilege the view of software systems as collections of interacting objects. The entire development process is built around an object-centric view of the application. In the analysis phase, the key abstractions are identified and encapsulated in objects. Design proceeds by breaking down these objects into lower level objects. Simplification – in modelling, design and testing – is achieved by partitioning the space of objects in an application into subsystems that are intended to be as self-contained as possible.

Our experience, however, is that when designing a framework, the focus of interest is on abstract interfaces and design patterns, not on concrete classes or components. Typically, the framework design process starts with the identification of design patterns which are then instantiated through the definition of abstract interfaces. In many cases, concrete components are also developed to support the abstractions introduced by the design patterns and their associated interfaces but this is always the last step in the design process and is conceptually subordinate to the definition of the design patterns and abstract interfaces. Once these have been identified and described, the definition of the concrete components follows naturally. This approach is illustrated for the case study of the AOCS Framework in chapter 8[8].

In accordance with the perceived priority of patterns and interfaces, design simplification in framework design must be achieved by subdividing the set of design patterns and abstract interfaces rather than the set of concrete classes. This insight is at the basis of the concept of framelet presented in the next chapter. In the

---

[8] This chapter introduces the case study. It is fairly self-contained and some readers may want to go over it before proceeding with the methodological material covered in this and the next chapters.

case of software frameworks, the traditional technique of achieving simplification by partitioning the space of components or classes will simply not work. Concrete components though an important part of frameworks and one of the three basic types of constructs that they offer, conceptually and from the point of view of the designer, take second place to design patterns and abstract interfaces.

There are fundamental reasons why design patterns and abstract interfaces naturally acquire greater salience in the framework development process. The development of individual applications is normally driven by very specific needs embodied in a formal set of requirements or use cases. The key problem in the analysis phase then becomes the identification of the behaviours subtended by the application specifications. Concrete classes become the focus of design attention because they are the most natural wrappers within which to encapsulate the behaviour implementations. The key problem in framework design, on the other hand, lies in the incorporation into its software of a sufficient degree of flexibility to allow it to be a basis for the instantiation of a vast number of different applications. Design patterns and abstract interfaces, rather than concrete components, are the ideal vehicles to model adaptability and this is why they acquire such prominence in the framework design process. This prominence must accordingly be reflected in a software development methodology for frameworks.

3.4.2 Support for Design Patterns and Hot-Spots

Developers of single applications often make use of design patterns to implement elegant and well-tested solutions to particular design problems. The design patterns they select are therefore *instantiated* in the application and normally appear in the application development process only in the architectural design phase. It is wise to record their use in the project documentation in order to make it easier to understand the application architecture or code but they remain otherwise at the margin of the software development process. It is not a coincidence that conventional methodologies hardly mention them and do not provide any guidance on how they should be modeled, described and used.

Design patterns play a very different and far more central role in the framework design process. A framework, as understood here, should *define* new design patterns that are useful for typical design problems arising in its domain. In fact, the design patterns are among the basic constructs that the framework exports and makes available to application developers. A methodology for software frameworks must consequently acknowledge their role and should provide conceptual tools for handling them.

Design patterns typically are introduced to model adaptability or, in the framework terminology, hot-spots. There is often a close relationship between design patterns and hot-spots. The disregard of design patterns by conventional methodologies is mirrored in their disregard for the concept of hot-spot. This disregard is of course justified by the fact that they are targeted at the development of closed applications where hot-spots seldom occur but it exposes another aspect of their inadequacy for handling framework development where instead the identification of hot-spots and their incorporation in the system architecture are essential steps in the software development process.

3.4.3 Iterative System Specification

As already remarked above, the development of an application is usually driven by some well-defined need that is external to the application development process. Conventional methodologies recognize this fact by constructing a software development process that is driven by a formal specification (expressed as a set of user requirements or of use cases) that, at least in theory, is supposed to be unchangeable. In practice, the results of the design process will often feed back into changes to the system specifications but this link is undesired and unintended, the mere unwelcome consequence of the inherent messiness of the software development process.

In chapter 5, we will argue that this approach to system specification is untenable in the case of frameworks which, since they represent a whole set of potential applications, have a higher level of abstraction. Specifying an artifact requires fully understanding the way it is to be used and, when the artifact in question has a high level of abstraction, a full understanding is difficult to achieve in advance of the artifact itself being developed. In a sense, the way a framework is to be used is *discovered* as the framework is designed. This creates a connection between the design phase and the specification process which conventional methodologies do not model but that that a framework development methodology must take into account by introducing *specification iterations*.

3.4.4 Design and Architecture

The software development process advocated by conventional methodologies is normally centered on an architectural definition phase that bridges the gap between the requirement analysis and coding phases. This gap is much wider in the case of frameworks than in that of single applications because of the greater complexity of the former. It is consequently sensible to interpose one or more additional stages between specification and coding. Here, this is done by introducing a distinction between architecture and design[9] [81].

This distinction is seldom made. There is general agreement (see for instance [50, 92, 25]) that an architecture is a description of the set of objects in a software system in terms of their external signatures (usually, their interfaces) and their mutual relationships (who is calling whom, who is acting as a server to whom, etc.). There is much less agreement over what constitutes a design. Sometimes de-

---

[9] The word "design" in this report is sometimes used to refer to the design phase in the framework development process as described in this section. At other times, it is instead used in the conventional sense to designated the process whereby the framework is defined. Context should make clear which usage is intended.

sign is simply seen as the *process* that results in an architecture. Sometimes it is regarded as identical to the architecture. This book will use this term to designate the organization of the software at the purely abstract level where the focus of attention is on finding abstract solutions offering the degree of adaptability required by the framework hot-spots rather than on defining objects and their relationships. On this view, *the architecture becomes an instantiation of a design.*

An example will clarify this distinction. Consider again the example of the Factory Method pattern. Its abstract representation was given in section 3.2. At that level, the design pattern represents a *design solution* to a design problem. Pattern instantiation takes place when the pattern is applied to a concrete architectural situation and results in the two classes of the pattern being mapped to specific classes and in other non-essential changes (like for instance having more than one factory method). At this point, the design pattern is instantiated and it represents an *architectural solution.*

A similar distinction between design and architecture can apply to an entire system. Indeed, analysis of the thought processes that lead to the definition of a software architecture shows that the designer passes through a design stage before arriving at the architectural stage. This is often ignored because standard development processes do not recognize an explicit design phase. This stage, however, becomes more evident in the case of frameworks whose design is considerably more complex than that of individual applications and where designers make a correspondingly larger investment in its definition.

One reference that takes a position similar to that advocated here is reference [9] that defines an *architectural style* as a class of architectures or an abstraction for a set of architectures. Its notion of architectural style thus resembles that of design as defined here. Most other authors, and all conventional methodologies, by contrast acknowledge only the architectural definition as an explicit software development phase. We instead give design the same prominence as architecture. This is in recognition of the difference between design and architecture and in keeping with the view expressed in the previous section that software reuse can occur both at architectural and design levels. Design reuse can occur only if design is accepted as a separate activity as is proposed here.

The recognition of a design level distinct from the architectural level is essential to making reuse possible in complex domains like the AOCS domain considered in the case study. In practice, in this type of domains, it is not possible to define a single reference architecture that can be entirely customized by plug-in components to generate all applications in the domain. There is simply too much variability in the domain for it to be captured by a fixed reference architecture. A software framework as it is understood by most framework practitioners – a reference architecture with a library of pluggable components – cannot be defined for such situations and pure architectural reuse becomes impossible. Reuse, however, is still possible at the design level as reuse of domain-specific design patterns. A separate design level was introduced to conceptualize this type of reuse.

There is a close relationship between the importance given here to design patterns and the recognition of a separate design phase since, as will be seen, the identification and definition of design patterns is the primary objective of the design phases (the framework and framelet concept definition phases, see sections 6.3 and 6.4). The definition of new domain-specific design patterns, or the adaptation of existing patterns to the framework domain, is one of the fundamental activities in the development of a framework. This activity is conceptually very different from that which is normally performed in an architectural definition phase (the definition of objects and their mutual relationships). It is this conceptual difference that advises against lumping both activities in a single development phase and that motivates the introduction of a new phase explicitly targeted at defining the framework design as expressed by the domain-specific design patterns.

Obviously, in practice, reuse will occur at all three levels defined in figure 3.3. In the AOCS Framework, for instance, code reuse is mediated by the provision of a large number of core and default components; architectural reuse results from the definition of component interfaces and component relationships; and design reuse is represented by the definition of exportable design patterns. Thus, our proposed approach, far from excluding or even de-emphasizing architectural and code reuse, simply adds to them a different and more abstract form of reuse that extends the applicability of the framework approach to complex domains.

3.4.5 A Methodology for Frameworks

The discussion in this section has identified four weaknesses of conventional methodologies with respect to framework development. Firstly, their emphasis on objects as the main actors in the development process is at odds with the more important role of design patterns and abstract interfaces that is inherent to frameworks. Secondly, they do not provide good support for modelling design patterns and hot-spots that are among the basic concepts of framework design. Thirdly, they see the software development process as driven by immutable system specifications and fail to recognize that in the case of frameworks there is a strong interplay extending throughout the design phase between specifications and architectural design. Lastly, conventional methodologies conflate in a single architectural phase activities that are conceptually different and that, in the case of framework development, are best allocated to different phases.

The next four chapters present methodological guidelines and concepts that are targeted at framework development and that try to remedy some at least of these weaknesses. Chapter 4 introduces the concepts of framelets and implementation cases that address design complexity as it arises in framework design. The following chapters consider the two main stages in the framework development process: the specification stage and the design stage. Finally, chapter 7 discusses the framework as seen from the point of view of the application developer. Wherever appropriate, the discussion is illustrated with examples from the AOCS case study.

It is hardly necessary to mention that a methodology for frameworks need not completely displace conventional methodologies. A framework contains a "conventional" part consisting of its components and their mutual relationships (the architecture *in senso stricto*). Their design and implementation can obviously be done following any conventional methodology. The methodological guidelines presented below should therefore be seen as complementary to conventional methodologies. They cover the parts of the framework development process that other methodologies cannot reach but they hand over to them when the development process re-enters their areas of competence.

The methodological solutions proposed in this book depend on four fundamental concepts – framelet, design pattern, implementation case, and functionality – that are specific to software frameworks and that will be unfamiliar to readers whose background is in application-level software development. It is possible to establish a parallel between these new concepts and some of the concepts used by traditional application development methodologies. The parallel arises from the fact that framework concepts can be seen as a transposition to a higher level of abstraction of familiar concepts in application design. Making this parallelism explicit can help provide insights into the concept of software framework and can aid appreciation of the methodology described in this report. The duality between framework and application concepts is presented in table 5. Readers may want to revisit this table as they proceed through the next four chapters.

Table 5. Duality of concepts between application concept (left-hand column) and framework concepts (right-hand column)

Application Development	Framework Development
Concrete Object ←→	*Abstract Interface/Des. Pattern*
The primary unit of design is the concrete object: the design process begins by identifying the concrete objects to be represented in the application and they remain the focus of design efforts.	The primary units of design are the interfaces and design patterns: the design process begins by identifying the abstract interfaces and the design patterns and they remain the focus of design efforts.
Subsystem ←→	*Framelet*
Design complexity is mastered by breaking up the application into subsystems. A subsystem is primarily a cluster of related objects.	Design complexity is mastered by breaking up the application into framelets. A framelet is primarily a cluster of related abstract interfaces and design patterns.
Use Case ←→	*Implementation Case*
A use case describes a potential usage of the application. Use cases can be used to specify the application, to check the application design and to describe its usage	An implementation case describes a potential usage of the framework. Implementation cases can be used to specify the framework, to check the framework design and to describe its usage.
User Requirement ←→	*Functionality*[a]
A user requirement describes an aspect of an application and can be mapped to the application architectural constructs.	A functionality describes an aspect of a framework and can be mapped to the framework architectural constructs.

(a) Functionalities are introduced in chapter 7

4 Framelets and Implementation Cases

Software methodologies are essentially means to master the complexity of the software development process. The complexity of framework development has at least two distinct aspects. *Quantitatively*, complexity stems from the sheer size of typical frameworks that may encompass hundreds of constructs – classes, design patterns, interfaces, hot-spots, etc – embedded in an often tangled web of interconnections and semantic relationships. *Qualitatively*, it arises from their high level of abstraction, itself a consequence of their attempt to model whole application domains. This chapter addresses both aspects of framework complexity. It introduces the *framelet* concept, that was developed specifically to address the quantitative complexity of framework development, and the *implementation case* concept that instead handles the qualitative side of framework complexity. Framelets are treated first.

In organizing this book, it was decided to have the discussion of methodological aspects precede the description of the architecture of the AOCS Framework. This order, which is otherwise the natural one, is unfortunate in respect of the introduction of framelets. This concepts – which had originally been proposed elsewhere [78, 79] – evolved to its present form during the development of the AOCS Framework as a direct result of the way in which this development process was organized. Understanding its genesis would probably help understand its characteristics and usefulness. Readers may therefore want to study chapter 8 that reconstructs the thought processes that led to the design of the AOCS Framework before reading this chapter. An analysis of how a software artefact comes into being is valuable because it may hint at rules of conduct that are applicable in general. The contour of the framelet concept was shaped precisely by such an analysis as applied to the fashioning of the AOCS Framework.

4.1 The Framelet Concept

One observation to emerge from the analysis of the AOCS Framework development process is that design patterns and abstract interfaces are the basis for framework design. Framelets formalize this insight and use it to provide a means to address the quantitative aspect of framework complexity. Framelets are introduced to simplify the design and the description of a framework by allowing it to be broken up into smaller and simpler entities.

A. Pasetti: Software Frameworks, LNCS 2231, pp. 49–67, 2002.

The need to tackle quantitative complexity is of course not restricted to frameworks. It also arises in the case of conventional application design where the classical solution is to divide the application into *subsystems* that partition the space of components making up the application. The subsystems are designed to be as self-contained as possible and to have minimal coupling. This makes it possible, to some extent, to develop them independently of each other thus reducing design complexity. This approach works well for individual applications that are just the sum of their components and, as initially proposed in [78, 79], the framelet concept mirrored it in simply subdividing the set of components offered by a frameworks into smaller subsets and treating framelets essentially as "small frameworks". However, frameworks consist of more than just components for they also include design patterns and abstract interfaces among their basic constructs. In a first modification, framelets were accordingly redefined to designate sets of logically related components *and* design patterns and interfaces [74]. In light of later experience, however, this view too had to be abandoned because, in the case of frameworks, components take second place to design patterns and interfaces that are the true foundations upon which frameworks are built. The behaviours implemented by the core components are normally implied by the design patterns and are therefore conceptually subordinate to them while default components are simply implementations of abstract interfaces.

Given these relative priorities among the constructs making up a framework, the transposition of the subsystem approach to software frameworks requires, at least in the design phase, a shift of focus from components to abstract interfaces and design patterns. Simplification in other words must be achieved by partitioning not the space of components, but that of design patterns and abstract interfaces. The framelet term was introduced to designate *non-overlapping groups of logically related design patterns and interfaces*. The two ways of partitioning the design space – through framelets and through subsystems – are schematically contrasted in figure 4.1. It is important to stress that there is no guarantee that framework components can be mapped to single framelets. This may be possible in some cases where use of the façade pattern will allow framelets to be neatly packaged as components. In general, however, a given component may implement several interfaces belonging to different framelets and may participate in several design patterns provided by different framelets. Such a component obviously cannot be assigned to any specific framelet. The framelet structure cannot therefore be imposed upon the space of components since components may straddle framelet boundaries. In the AOCS Framework case study, for instance, virtually all components offered by the framework intersect multiple framelets simultaneously. This point is illustrated in the case of the AOCS Framework in section 8.10. Conversely, it is normally possible to partition the space of abstract interfaces and design patterns into homologous subsets because the abstract interfaces exist to support the variability and adaptability introduced by the design patterns and therefore abstract interfaces can be unambiguously associated to design patterns.

Framelets divide the set of design patterns into subsets that represent self-contained design solutions. In order to correctly understand the concept of framelets, it is necessary to be very clear about *which* set of design patterns is thus sub-divided. A framework proposes a generic solution for a group of related applica-tions. One important (perhaps the most important) part of this solution consists of a number of domain-specific design patterns that are offered to application deve-lopers to help them solve design problems that arise in the framework domain. It is precisely this set of domain-specific design patterns offered by the framework that is partitioned by the framelets (see also the discussion in section 8.8). Diffe-rent frameworks, even if they address the same target domain, will probably come up with different sets of domain-specific design patterns. There is therefore not-hing intrinsic or inevitable about the relationship between framelets and a parti-cular domain. The analogy with subsystems can again help clarify this point. Con-sider two designers who are given the same set of application requirements. The design solutions they propose, in terms of the classes they use to model the appli-cation abstractions and the way they group them into subsystems, will probably be very different. Similarly, two different framework design teams are likely to pro-pose two different sets of design patterns – and hence two different sets of frame-lets – to solve the same domain problems.

Finally, the relationship of hot-spots to framelets should be considered. Hot-spots are not considered as primitive constructs in a framework because they are reducible to design patterns and abstract interfaces but it is interesting to note that, precisely for this reason, hot-spots can be assigned to framelets. Framelets, there-fore, can also be defined as a partitioning of a framework into subsets of logically related abstract interfaces, design patterns and hot-spots. In practice, logical relati-onship means that each framelet addresses a specific design problem in the fra-mework. Framelets in other words become *units of design* with the framework being obtained by combining the framelets.

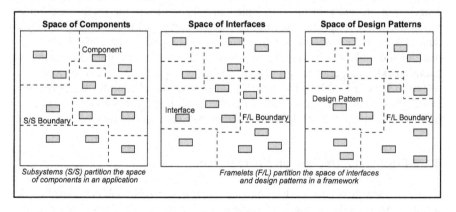

Fig. 4.1. How framelets and subsystems partition the space of software constructs

4.1.1 Implications for the Design Process

Adoption of a framelet-based approach to design a framework has profound implications on the framework design process. As mentioned in the previous chapter, conventional design methodologies aimed at individual applications begin the design process with the identification of the basic concrete objects in an application. These are then subdivided into simpler objects thus resulting in a system that is hierarchical in the sense that its objects and classes are nested within each other. This way of proceeding is the natural one when complexity is mastered by partitioning the space of components or modules. If, however, complexity management is done at the level of abstract interfaces and design patterns, as is the case with framelets, then attention must shift to the interfaces and the patterns. The design process must start by identifying abstract functionalities that are then encapsulated in abstract interfaces, and points of framework adaptability to which design patterns are fitted. Concrete classes and the components instantiated from them are introduced only at a later stage as implementation vehicles for the interfaces and the patterns. The resulting architecture will again be hierarchical but in the different sense of being based on a hierarchical class tree with abstract classes and generic design patterns at the top of the tree and concrete classes and domain specific design patterns at the bottom. The design process we are advocating takes this into account and, as will be seen in a later chapter, one of the objectives of the first phase of the process – the framework conceptual design phase – is precisely to identify the major design patterns, hot-spots and abstract interfaces in the framework domain.

After having emphasized the parallelism between framelets and subsystems, an important remark is in order. Framelets are introduced to simplify the *design* process but are unlikely to be of much assistance during the *framework implementation* phase. Implementation is largely concerned with the coding of individual classes and classes cannot be assigned to framelets. Thus, whereas it is in principle possible to distribute the framework design effort among different design teams working in parallel with each team being responsible for one or more framelets, it is not possible to preserve the same work distribution for the implementation phase. This is an important difference with respect to subsystems that can be used to simplify both design and implementation of individual applications.

4.1.2 Framelets, Design, and Architecture

One of the key principles of the approach advocated in this book is the explicit distinction between design and architecture (see section 3.4.4). As will be seen in a later chapter, framework development begins at the design level with the analysis of the software at the purely abstract level to find abstract solutions offering the degree of adaptability required by the framework hot-spots. This is followed by the traditional architectural definition phase where components, their interfaces, and their mutual relationships are studied. Framelets are introduced to organize the design phase because it is in the design phase that notions like hot-spot, abstract interface and design patterns play a central role. As the framework development proceeds into the architectural phase, framelet borders will become increasing blurred as framelets are integrated to form a unified and coherent architecture. At that point, other means of simplifying software analysis will have to be used. These are not discussed in this report because the problem of dividing an architecture – namely a collection of cooperating components – into subsystems is regarded as well-known and is addressed in countless technical references.

One criticism that might be made to the framelet concept as the fundamental tool to shape the framework design is that framelets, in addressing *local* design problems, neglect the large-scale structure of the framework. Some authors (for instance [20]) distinguish architectural patterns – that address system-wide architectural issues – from design patterns – that address specific design problems arising in particular locations of a system. Well-known examples of architectural patterns (or architectural styles to use still another terminology) are the layered pattern, the pipe-and-filter pattern, and the object pattern [92]. All of these patterns prescribe how the architecture should be organized in-the-large. It might be argued that the framelet approach does not provide any way of addressing this type of system-wide issues. This argument however is wrong and arises from a confusion between the design and architectural levels.

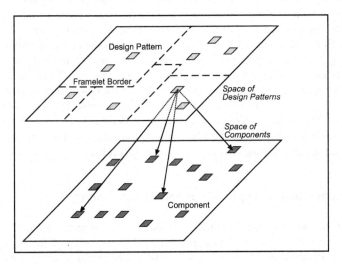

Fig. 4.2. How design patterns can have a global action at architectural level

At design level, framelets propose specific design solutions to specific design problems and may appear to be "local" in character. These solutions, however, when carried over to the architectural level might well affect the entire framework or an entire application instantiated from the framework. Consider for instance the mode management framelet in the AOCS Framework case study (see chapter 12).

At design level, this framelet proposes a mechanism – a domain-specific design patterns – to let components in the AOCS applications vary their behaviour in response to changes in their environment. At this level, this pattern is indeed local in the sense that it is independent of the patterns provided by other framelets. The framelet simply covers one small chunk of the total design space. At architectural level, however, the pattern defined by the framelet pervades the entire framework as it is applied to any component that needs to exhibit mode-dependent behaviour. At architectural level, therefore, the operational mode pattern has become system-wide because it affects the entire architecture. This situation is illustrated in figure 4.2 where a single design pattern is shown to affect components that are spread over the entire application.

One difficult question that arises at this point is that of the mutual interactions of the patterns when they are instantiated in the architecture. If, at design level, the design solutions provided by the various framelets are self-standing and self-contained, at architectural level they become intermingled and the possibility arises that incompatibilities might arise. This is precisely the case because solutions that are "local" in the design space may become "global" or "system-wide" in the architectural space. This risk of conflicts is of course not specific to the framelet approach. It arises whenever a solution identified at one level of abstraction has to be carried over and implemented at a lower level of abstractions. Conventional methodologies simplify the definition of the architecture of individual applications by dividing the application into subsystems. Efforts are made to ensure that the subsystems are as decoupled as possible so that the designer may consider them in isolation from each other. However, when the subsystems are integrated during the coding and implementation phases unexpected interactions may emerge that undermine the architectural decomposition into subsystems. The problem of interference between the design solutions of different framelets is the same but at a higher level of abstraction as it arises in the transition from design to architectural definition rather than in that from architectural definition to implementation. In both cases, the best protection against the problem arising is the experience of the software designers, and the only remedy when the problem does occur is iteration.

4.1.3 Framelets and Aspect-Oriented Programming

Framelets do not partition the space of components. Aspect Oriented Programming [57] may offer a way of conceptualizing the way in which the same component can participate to different framelets. In its terminology, a feature of a software system that affects, could potentially affect, or requires the cooperation of multiple classes in the system is said to *cross-cut* the classes. To state it more generically, a feature or behaviour which cannot be readily encapsulated in one modular unit (subroutine, procedure, module, class, etc.) offered by the programming language or design technique being used is said to cross-cut the modular units. Such cross-cutting features are referred to as *aspects*. Framelets can be seen as providing the definition, via their design patterns and abstract interfaces, of how components should implement certain aspects of their overall behaviour. In this sense, framelets become an *Aspect-Oriented Design* technique for frameworks.

An example from the AOCS Framework case study will help clarify the relationship between framelets and aspect oriented programming. Consider the telemetry management framelet (see chapter 21 or section 8.3). This framelet introduces the concept of *telemeterable component*. Telemeterable components are components whose state can potentially be included in the telemetry stream. They are characterized by the interface `Telemeterable` for which they must provide an implementation. In general, many components in the AOCS Framework and in an application instantiated from the framework would implement interface `Telemeterable`. Using the terminology of aspect oriented programming, it might therefore be said that being telemeterable (as defined by implementing the `Telemeterable` interface) is one aspect of the framework components. This position is further justified by the fact that the implementation of the telemeterable interface is very similar across classes. There is much duplication of effort involved in implementing `Telemeterable` for different classes although this duplication cannot be readily captured by conventional programming means (it cannot, for instance, be eliminated through the use of templates, or by delegation to plug-in components, or by making recourse to subroutine libraries). The type of duplication introduced by the implementation of `Telemeterable` is in fact rather similar to the type of duplication introduced by the need to make a piece of software safe for multi-threading operation and synchronization is the classical example of an aspect. Also, if the `Telemeterable` interface changes, there will be several places throughout the implementation which will have to be changed to accommodate the changes in the `Telemeterable` interface.

As another example, consider again the mode management design pattern of chapter 12. The pattern is proposed to make the behaviour of components operational mode-dependent. It is used in several places in the frameworks and would also be used by developers of applications within the framework domain. Inspection of the code where the pattern is used shows again considerable similarities of a kind that cannot be modeled with conventional programming constructs. Mode dependency can therefore also be seen as an aspect that pertains to some components in an AOCS applications.

Both the telemeterable and the mode management examples point to the existence of large-scale similarities in the architecture of a framework that cannot be described in conventional terms. There would be clearly much to be gained from discovering and factoring out those similarities into a separate, single implementation. Using again the terminology of aspect oriented programming, what is needed is to find an aspect language to express the non-conventional framelet constructs (the abstract interfaces and the design patterns) and to develop tools that allow this language to be weaved to the conventional languages in which the framework components are developed. The availability of such languages and tools would extend the usefulness of framelets from the design to the implementation phase. This line of research was not pursued so far. It represents one of the directions in which the work done here could be extended as discussed in chapter 1. In particular, to the extent that aspect-based tools could automate the process whereby user components are made to participate to the framework (e.g. by making them implement framework interfaces or by making them support framework design patterns), they would help create development environments that allow application specialists to directly assemble their applications using tools and components supplied by a framework.

4.1.4 Framelet Features

There are three features of framelets that follow from how they are defined in the previous sections. Firstly, framelets typically have a *small size.* Frameworks are normally made up of a large number (sometimes running into several hundreds) of interrelated and cooperating constructs. Framelets are designed to be small to increase manageability. In the AOCS Framework case study, a framelet typically includes up to three design patterns, around 4-6 abstract interfaces and between 5 and 10 hot-spots (of which 2 or 3 would be major hot-spots). Note that it is not always possible to say how many classes belong to a framelet because, as explained in previous sub-sections, the framelets do not necessarily partition to space of classes/components.

Secondly, framelets make *no assumptions of execution control.* Frameworks often assume that they have control of the application execution and are consequently difficult to integrate with other software [66, 20]. Framelets make no such assumption and are designed to be amenable to integration with each other and with other software. Indeed, the full framework is obtained by combining several individual framelets. This feature also makes a framelet-based framework easier to compose with other frameworks. More generally, a framelet approach simplifies the extension of a framework since it makes it possible to add new functionalities to the framework by adding new framelets.

Thirdly, framelets are *self-standing.* Although frameworks are intended to be integrated together to build a full frameworks, they are also designed to be self-standing and to be in principle usable in isolation from the other framelets. Framelets can be used independently of each other because they address specific design problems and provide complete solutions to them. To the extent that the same problem can arise in contexts outside the framework domain, then the framelet can be reused independently of the framework. For an example from the AOCS Framework case study, consider the case of the telemetry management framelet (see chapter 21). This framelet proposes a solution to the problem of managing the stream of housekeeping data that a satellite must periodically send to the ground station. This framelet is self-standing because the design patterns and abstract interfaces it defines could also be used in the context of, say, industrial control systems to manage the flow of data to the operator or in any other context where there is a problem of managing a data-collection facility. This last feature in a sense implies a rule for deciding of the size of a framelet: a framelet should be as small as possible while retaining sufficient functionality to be independently usable.

It is noteworthy that all the above features contribute to simplifying design and justify the claim that framelets help reduce the complexity of the framework design process.

4.1.5 Framelet Constructs

A framework as conceptualized here consists of abstract interfaces, design patterns and components. Since in a framelet-based approach, the framework must be reducible to its framelets, these constructs must ultimately be derived from constructs exported by the framelets. From the discussion in the first part of this chapter, it should be clear that both abstract interfaces and design patterns can be unambiguously assigned to framelets. Components instead pose a problem because framelets do not partition the space of components in the framework. However, the framework often offers default implementations of abstract interfaces. Since the interfaces belong to specific framelets, it can be said that the framelets also offer *interface implementations*. In practice, these implementations will be packaged as components but the same component might implement many different interfaces coming from different framelets and therefore cannot be mapped to any particular framelet.

To exemplify, suppose that framelet F_A exports interface A and framelet F_B exports interface B. Consider then a component C exported by the framework that implements both interfaces A and B. Clearly, this component cannot belong *in toto* to either framelet F_A or F_B, however the implementation of interface A offered by C can belong to framelet A and similarly the implementation of interface B offered by C can belong to framelet B. In the language of a previous section, one could also say that framelets export *aspects of components*.

Thus, strictly speaking, a framelet exports the following constructs:

- Abstract Interfaces
- Design Patterns
- Interface Implementations built into components

Very often, of all the interfaces implemented by a component there is one that is dominant in the sense that it is the one that occasioned the creation of the component. In that case, and with an abuse of language, expressions like "component exported by a framelet" will be used for reasons of convenience. It should however be borne in mind that they are improper and that they refer to the implementation of some interface packaged in the component.

4.1.6 Framelet Heuristics

The breaking up of the framework into framelets is one of the crucial steps in the early part of framework design. In importance – and in difficulty – this task is comparable to that of identifying key application abstractions[10] and mapping them to objects in conventional application design. Our experience suggests five guidelines to facilitate this task presented in the following five subsections:

- *Mapping clusters of related requirements to framelets.* In the AOCS project, the framework was derived from an analysis of a set of AOCS applications each described by a set of user requirements. User requirements are often organized in groups of related requirements. One useful heuristic for identifying framelets is to find the requirement groupings that recur in many applications. In practice, this can often be done by inspecting the table of contents of several user requirement documents and identifying recurring sections. Many methodologies for application development suggest that potential objects can be recognized by underlining often-recurring nouns in requirement specifications [6]. The heuristic proposed here to isolate potential framelets is similar but operates at a higher level of abstraction.
- *Building framelets around single or related hot-spots.* The identification of hot spots is one of the early tasks in the framework design process. A major hot-spot can be the core around which a framelet is built.
- *Building framelets around design patterns.* Design patterns are among the building blocks of frameworks. Major design patterns are identified early in the framework design process to solve localized design problems. They typically involve only a handful of classes and abstract interfaces and usually make no assumptions of about execution control. They can therefore serve as the basis for a framelet. Note that this heuristic is related to the previous one because design patterns are often introduced to model hot-spot variability.
- *Mapping tasks to framelets.* Applications, and in particular real-time applications, are often organized around tasks representing separate threads of control. In many cases, tasks are created statically and therefore their number and function is defined at design time. Tasks typically encapsulate self-contained functionalities and have well-defined boundaries with each other. They are thus good candidates for framelets and one natural guideline in framelet identification is to look for typical application tasks and to map them to framelets. Mapping framelets to tasks ensures that the framelets are functionally decoupled (thus facilitating their design) and that they do not make any assumptions about the control of their own execution which is explicitly delegated to a scheduler (thus facilitating their integration into a single framework).
- *Mapping abstract use cases to framelets.* Abstract use cases are introduced in [68]. They are found by searching for patterns in a large number of use cases for applications in the framework domain. They embody abstract forms of behaviour that are common to many applications in the domain. Reference [68] proposes them as a way of identifying abstract classes in a framework but they

---

[10] In order to avoid any misunderstanding it may be well to point out that *application* abstractions are different from *framework* abstractions. The former typically model objects, the latter typically model variability and adaptability.

could also be used to identify framelets since a framelet should ideally encapsulate one particular form of behaviour variability.

In practice, an economical design will have several heuristics point to identification of the same framelet. In the AOCS Framework case study, for instance, all but one heuristics indicate that telecommand handling should be covered by a dedicated framelet. This convergence of criteria enhances confidence in the decision to treat telecommand handling as a single framelet. Each project may select its own set of heuristic rules for framelet identification. A table can then be set up where tentative framelets are listed and ticked against applicable heuristics. See table 6 for an example. Clearly, when there are framelets to which only few heuristics apply, thought may be given to merging them with other framelets or dropping them altogether.

Table 6. Sample table for identifying framelets

	Heuristic 1	Heuristic 2	Heuristic 3	. . . .
Tentative f/l 1	✓		✓	. . . .
Tentative f/l 2	✓	✓		. . . .
Tentative f/l 3			✓	. . . .
Tentative f/l 4		✓	✓	. . . .
. . . .	. . . .	. . . .	. . . .	. . . .

4.1.7 Framelets in the AOCS Framework

The AOCS Framework is organized as a set of framelets and provides examples of usage of all but one framelet heuristics (the last heuristic based on abstract use cases could not be used because AOCS systems are not normally described by use cases):

– Of the 13 framelets proposed for the AOCS Framework, six – the *manoeuvre management, controller management, telemetry, telecommand, failure detection*, and *failure recovery framelets* – correspond to clusters of requirements that will be found in virtually all user requirement documents for AOCS applications.
– The *telemetry* and *telecommand framelets* map directly to tasks in the AOCS software which normally devolves telemetry and telecommand management to dedicated tasks.
– The *data processing* and *controller framelets* do the same as their function is to provide components to implement attitude and orbit control algorithms and the implementation of these algorithms is usually allocated to a dedicated task.
– The structure of AOCS units (sensors and actuators) is very application-specific and therefore unit management is a natural hot-spot in an AOCS Framework. The AOCS *unit framelet* was built around this hot-spot. It consists of an abstract interface that encapsulates the generic operations that can be performed

on any AOCS sensor or actuator but leaves the implementation open to individual applications.
- Failure detection, failure recovery and manoeuvre management are also highly application-specific and give rise to hot-spots in the AOCS Framework which then became the basis for three framelets. In their case, however, it was possible to identify some application-independent behaviour that was packaged as components exported by the framelets to end-applications.
- The algorithms used to process the AOCS data are another obvious hot-spot which was encapsulated in the *data processing framelet*.
- Most components in an AOCS software are required to adapt their behaviour in response to changes in their environment. This effect is usually achieved by endowing them with mode-dependent behaviour: components are given control over several algorithms – one for each mode – and employ application-specific rules to select the one to be executed at any given time. The selection is a function of the state of other components. A design pattern was devised to provide components with mode-dependent behaviour and became the basis of the *operational mode framelet*.
- Monitoring of component properties is another common task in AOCS systems and for it, too, a generic design pattern was devised that became the basis of the *object monitoring framelet*.

One of the key properties of framelets is that they should be self-standing – reusable independently of each other – and potentially usable in contexts or domains different from those for which they were initially proposed. The presentation of the AOCS framelets in the second part of the book includes a discussion of their reusability outside the AOCS domain. It is noteworthy that of 13 framelets only 1 (the *Unit Framelet*) is entirely AOCS-specific. All the others could in principle find application in embedded control systems in general and some have an even wider applicability. However, in most case, the AOCS framelets do not carry over to the architectural level (i.e. framelets do not partition the space of components). Hence, reuse of framelets in domains other than the AOCS can usually be done at design level only.

Thus, the AOCS specificity of the framework described in this report lies less in its constituent parts – the individual framelets – than in their composition. The fact that the AOCS design could be broken up into smaller entities that are not AOCS-specific is regarded as evidence of the high quality of the design choices made in the AOCS Framework project and as a vindication of the framelet approach.

4.1.8 Related Approaches

Framelets are proposed as a means to simplify the *design* definition as opposed to the *architecture* definition of a framework. Other simplification approaches tend to be more at the architectural level. [43] for instance proposes the concepts of *class families* and *class teams*. A class family is a set of classes that share the same abstract interface or base class. A class teams is a set of cooperating class families. Such concepts are undoubtedly useful at architectural level but they are not very helpful at design level where class trees have not yet been built and where the emphasis is rather on design patterns and hot-spots than on concrete classes.

The approaches that come closest to the framelet-based one advocated here are those that rely on a division of the domain into subdomains for which frameworks (or parts of frameworks) are then developed in isolation from each other. One such author is [4] who decomposes a framework into subsystems where one subsystem implements the functionalities of a particular subdomain of the framework domain. Framelets can be seen as solutions for design problems arising in subdomains although the view of reference [4] of the framework as a variance-free architecture plus a set of pre-defined plug-in components suggests that the subsystems are just collections of related components.

Similarities might also exist with the FORM method of [55, 56]. FORM is based on the concept of features that describe "distinctively identifiable functional abstractions that must be implemented, tested, delivered and maintained" [55, p. 144]. Bundles of related features are similar to framelets and subdivision of the feature space might lead to a design simplification similar to that achievable with framelets. However, neither of the above references pursue this possibility. This is probably because they make no formal distinction between a design and an architectural phase and this distinction is essential to appreciate the role of framelets (or other framelet-like constructs) since, as soon as one enters the architectural level, simplification is best achieved in the traditional manner by acting on class trees.

Reference [20] proposes that the architectural definition should begin with the identification of so-called *archetypes*. Archetypes capture key domain abstractions but, although they usually do not correspond to directly identifiable entities in the target applications, they are encapsulated in objects. However, its author then mentions (pag. 60) that small groups of related archetypes may form system-specific patterns applicable in many locations with the systems. This notion may be close in spirit to the framelet approach if the small groups of archetypes also include the conceptual relationships between them to form design patterns that address specific design problems in the domain.

Reference [13] presents still another approach to design simplification based on the concept of "collaboration" or "class layer" to encapsulate an aspect of a framework. This approach is, at the same time, more sophisticated and more restrictive than that advocated here. It is more sophisticated because the combination of the layers is formalized and framework instantiation can in principle be completely automated. It is more restrictive because the framework is essentially seen as a set of classes of which some are abstract and others are concrete and implement behaviours that are needed by all applications in the framework domain. This type of conceptualization restricts the scope of framework since many domains – the AOCS domain considered in the case study being an example – cannot be easily modelled by sets of concrete classes.

4.2 The Implementation Case Concept

Frameworks are largely made up of abstract constructs. As their design proceeds, there is a natural need to check the adequacy of the constructs that are proposed – abstract interfaces, components and design patterns – without waiting for the prototyping phase. The concept of implementation case was introduced [74] to cover this need and thus to address framework complexity from its qualitative angle.

A framework is a tool to help developers rapidly build an application within the framework domain. An implementation case describes an aspect of this application instantiation process. It specifies how a component, an architectural feature, or a functionality for an application in the framework domain can be implemented using the constructs offered by the framework. An example from the AOCS Framework case study will clarify this definition. The AOCS Framework is a tool to assist the development of AOCS applications. An important functionality of any AOCS application is the processing of sensor information to reconstruct the satellite attitude. Accordingly, the following implementation case can be formulated: "build a component that uses sensor measurements to produce an estimate of the satellite attitude". This implementation case is worked out by showing how the constructs offered by the framework are combined to build the required component. Thus, implementation cases define an objective for a localized instantiation action. They are said to be *worked out* when they are accompanied by a description of how their instantiation objective can be achieved using the framework. When the framework design is completed, implementation cases can be worked out in detail essentially resulting in cookbook-style recipes for using the framework [60]. When the design is still under way, only partial working out of the implementation case is possible since not all the framework constructs are yet available or finalized. However, even at early design stages, going through the implementation cases remains very useful because the exercise can reveal shortcomings in the already defined constructs and can point towards constructs that are still needed.

The term "implementation case" was coined by analogy with the term "use case" as employed by some methodologies for application development. A use case describes the way an application is intended to be used. Use cases cannot be defined for a framework because a framework is *not* a working application and it is not *used* in the same sense in which an application is used. An *implementation case* is its equivalent in the sense that a framework is a tool to help implement applications and implementation cases describe how a feature of an application can be implemented.

Implementation cases are also related to the Software Architecture Analysis Method (SAAM) scenarios [9]. SAAM scenarios can act as tools to measure the adaptability of an application to future changes. A SAAM scenario describes a hypothetical change in the application specification and considers the ease with which the application design and implementation can be modified to meet the new specifications. An implementation case is similar in that it describes a scenario for adapting an architecture to a newly introduced set of requirements. The difference is that SAAM scenarios are typically targeted at individual applications which are not specifically designed to be adapted whereas implementation cases are targeted at frameworks that exist precisely to be adapted.

Still another related idea can be found in [20] which introduces the concept of *profile* as a means to evaluate the quality attributes of a software application. A profile is a set of scenarios. Use cases are a special type of scenarios which describe how the application can be used and are useful to evaluate the application from a functional point of view. Reference [20] proposes other kinds of scenarios that offer other perspectives on the application. *Hazard scenarios,* for instance, can be used to evaluate safety and *change scenarios* can be used to evaluate maintainability (pag. 83). One way of assessing the ability of an architecture to achieve its stated levels of performance both with respect to its functional and non-functional requirements is to perform *simulations* of the scenarios (pag. 95). This involves executing the system at a level sufficient to allow the assessment to be performed. Implementation cases can be seen as a form of simulation in the sense of [20] but of course their goal is specific to frameworks and differs from the goals considered in that reference.

A more tentative connection may also be established between implementation cases and extreme programming [14]. The framework approach is obviously alien to the philosophy of extreme programming (XP) with its rejection of reuse as a legitimate objective of the software development process and its minimalist attitude to design. However, adoption of implementation cases furthers some of the key principles advocated by XP such as *rapid feedback,* since they allow very short turn-around times between design and implementation, and *concrete experiments,* since they make design experiments possible at an early stage in the development process. To the extent that implementation cases resemble minimal self-contained implementations, their usage promotes practices like *small releases* and *continuous integration* that are key to the XP paradigms. Finally, and again in line with the XP philosophy, implementation cases shifts the emphasis from design to testing (or, more precisely, to the testing of the design). The relationship between implementation cases and the XP style is however only analogical because XP is concerned with the rapid prototyping of applications whereas implementation cases are concerned with the rapid design of frameworks. One could perhaps describe implementation cases as an "extreme design" approach to framework development.

4.2.1 The Three Roles of Implementation Cases

Implementation cases can play at least three roles in the framework design process. In an early phase, they help check the adequacy of the design ideas. Typically, whenever a new major construct is introduced, its effectiveness should be tested by working out an implementation case that uses it. Where necessary, new implementation cases must be introduced to cover the functionalities introduced by the newly defined construct. In our experience, this process of refinement of implementation cases is the single most important source of changes in the framelet design in the first part of a project and it can probably replace at least one iteration cycle in the design process. Thus used, implementation cases address the qualitative aspect of framework complexity because they force designers to think about the reification of the abstractions they are creating while at the same time giving them the opportunity to check the adequacy of these abstractions. Implementation cases are therefore a *design testing tool*: they offer a means of continuously testing the design of the framework as it evolves during the framework development process and well in advance of the coding stage.

Use cases are primarily a means to specify an application. Implementation cases can be used for the same purpose since a sufficient number of them could cover all the functions of a framework and could thus be a way of specifying it. The acceptance test for the framework then becomes its ability to implement all the components and architectural features described in the implementation cases. The ease with which this can be done is a measure of its quality: a well-designed framework offers abstractions and components that let users quickly and naturally assemble the constructs required by the implementation cases.

Finally, implementation cases can play a further role. At the end of the framework development process, after they have been fully worked out in parallel to the framework design, they are available as commented pseudo-code ready for inclusion in the framework user manual. Implementation cases then become cookbook-type recipes showing how framework constructs can be used to develop small applications or fragments of applications.

4.2.2 Implementation Case Scenarios and Extensions

Very often, to each use case several *use case scenarios* are associated with each use case scenario particularizing the use case to a specific and concrete set of circumstances. The equivalent concept of *implementation case scenario* can also be introduced with the same function to make an implementation case more concrete. Consider for instance a generic implementation case formulated as follows: build a component representing a generic closed-loop controller. The following implementation case scenarios could be associated to it:

- *Build a component representing an attitude controller*: This scenario particularizes the generic implementation case to build a component to implement a controller for the satellite attitude.
- *Build a component representing an orbit controller*: This scenario particularizes the generic implementation case to build a component to implement a controller for the satellite orbit.
- *Build a component representing a wheel-speed controller*: This scenario particularizes the generic implementation case to build a component to implement a controller for the speed of one of the satellite's reaction wheels.

If implementation cases are used only as a way of continuously monitoring the adequacy of a proposed design, then generic implementation cases are probably sufficient. Scenarios may become advantageous when implementation cases are used in the framework user manual since in that case more concrete examples are required than can be provided by the generic implementation cases.

Use cases can be in an *extension relationship* with respect to each other. Use case A is said to extend use case B when A includes the actions foreseen by B. An equivalent relationship of extension can be defined for implementation cases with implementation case A being said to extend implementation case B when A includes the instantiation and implementation actions foreseen by B. In the case of the example implementation cases mentioned above – build a component representing a generic closed-loop controller – one of its possible extensions could be:

– *Build a component representing a 3-axis attitude controller*: A 3-axis attitude controller in a satellite is normally realized as three parallel single-axis controllers. The base implementation case can be used to generate a single-axis controller and hence this implementation case is an extension of it.

4.2.3 Description of Implementation Cases

We are not proposing any special formalism to represent and model implementation cases. Implementation cases are best described in a mixture of informal language and pseudo-code with UML diagrams to assist the description. Lack of a dedicated formalism however does not mean arbitrariness and we recommend that implementation cases be described in an informal but systematic manner. For each implementation case, the following information should be provided:

– Implementation case objective
– Objective description
– Framelets involved in the implementation case
– Framework constructs involved in the implementation case
– Description of how the implementation case is worked out

Note that in the approach proposed here implementation cases are defined incrementally during the design process. Hence, the information items listed above are not all supplied at the same time. They are instead provided gradually as the framework design matures and the constructs for working out the implementation cases become available. Initially, only the objective and its description are provided. Other fields are filled in gradually as the framework design is refined. The implementation description section, in particular, will be expanded and made more detailed as the constructs for working out the implementation case become available. Eventually, the implementation description section will contain commented pseudo-code showing how the implementation case objective can be met by combining and adapting framework constructs. The degree of maturity of implementation case description can be used as a measure of the maturity of the framework design.

It is perhaps useful to offer two examples built around the AOCS Framework case study. The first example in table 7 deals with the implementation of a common kind of attitude manoeuvres performed by satellite. A "slew" is simply a rotation of the main body of the satellite around a fixed axis. The table describes an implementation case for the AOCS Framework at a level of detail that is adequate to the early phase of the framelet design process. As the framelet design proceeded, the implementation case description section should be made more detailed. In a later iteration, for instance, the description should include hints on how the attitude slew subclass should set up a link with the attitude control set points. Eventually, pseudo-code should replace the implementation description. The pseudo-code is not shown as understanding it would require more background on the AOCS Framework than can be provided here.

Table 7. First implementation case example

Objective	Implement an attitude slew manoeuvre.
Description	Attitude slews are common types of manoeuvres performed by satellites. The AOCS Framework encapsulates manoeuvres in components. This implementation case shows how to build a component encapsulating an attitude slew manoeuvre.
Framelets	Manoeuvre Management Framelet
Constructs	Manoeuvre Abstract Class (exported from Manoeuvre Management Framelet)
Hot-Spots	Manoeuvre Hot-Spot (exposed by Manoeuvre Management Framelet)
Related ICs	None
Implementation	– Subclass class `Manoeuvre` to build an attitude slew manoeuvre – Implement the attitude slew subclass to update the set-point of the attitude controller according to the profile prescribed by the attitude slew

Table 8 shows a second example of an implementation case for the AOCS Framework. The level of description is the same as in the previous example. In this case, its further working out during later stages of the design process would include specifying how the attitude manoeuvre component is to be loaded into the manoeuvre manager. Note furthermore that this implementation case *extends* the previous one in the sense that it uses its output and logically follows it up.

In the case study part of the book, several implementation cases for the AOCS Framework are presented. The level of detail of the presentation is somewhat constrained by editorial considerations and space limitations but they should be sufficient to give a better idea of how implementation cases are constructed and used

Table 8. Second implementation case example

Objective	Build a telecommand to perform an attitude slew manoeuvre.
Description	Attitude slews are normally started by a ground command (telecommand). This implementation case shows how to build a telecommand to perform an attitude slew manoeuvre. It is assumed that the attitude slew manoeuvre is encapsulated in the component built in the implementation case of table 7.
Framelets	Telecommand Framelet Manoeuvre Management Framelet
Constructs	`Telecommand` Interface (exported from Telecommand Framelet) `Manoeuvre` Abstract Class (exported from Manoeuvre Management Framelet) `ManoeuvreManager` Component (exported from Manoeuvre Management Framelet)
Hot-Spots	Telecommand Hot-Spot (exposed by Telecommand Framelet)
Related ICs	Attitude Slew Manoeuvre Implementation Case (this implementation case uses the component built in the attitude slew manoeuvre implementation case of table 7)
Implementation	– Subclass class `Telecommand` to build a manoeuvre telecommand – Implement the manoeuvre telecommand subclass to load an attitude manoeuvre component into the manoeuvre manager.

5 Framework Specification

In the specification phase, a formal description of the framework domain must be derived to guide and constrain the development effort by defining the functionalities to be offered by the framework. The investigation of ways to construct and express this type of formal descriptions is the object of a dedicated academic discipline – requirements engineering – and it is certainly not the intention here to reproduce its results. Instead we would like to offer some considerations borne out of our experience that pertain to the problem of specifying a framework for a complex domain, such as that of satellite control systems considered in the case study, together with an argument in favour of implementation cases as a specification vehicle for software frameworks.

5.1 How Important Is Specification?

The framework specification process depends crucially on the source from which the specification information are derived. Two possibilities can be distinguished in this respect:

– The framework specifications are derived *a priori* from an analysis of the application domain, and,
– The framework specifications are derived *a posteriori* from an analysis of a set of concrete applications by, for instance, inspecting their code and architecture and identifying commonalities and recurrent patterns.

The latter procedure seems to be the most common in practice [68] and we believe that it is the only one to be really viable in the case of complex domains. This is because, for this kind of domains, framework development is only possible if the domain is well-understood by domain experts that participate in the framework development. The objective of the specification phase should then be less to understand, or gather information about, the domain itself than to formalize existing knowledge and express it in a form that is suitable to drive the framework development. In practice, this will only be possible if several individual applications in that domain have already been developed.

Framework specification *a priori* and in the absence of individual applications is regarded as unrealistic, at least for domains with a high degree of complexity. It is probably equally unrealistic to expect that domain expertise can be transferred to a development team through domain analysis or through intensive exchanges

A. Pasetti: Software Frameworks, LNCS 2231, pp. 69-77, 2002.
© Springer-Verlag Berlin Heidelberg 2002

with domain experts. Our experience is that framework development for a complex domain is only possible with the direct and prolonged participation of domain experts. These can be either software engineers who have already developed several applications in the domain, or domain engineers with a sufficient understanding of software issues to contribute to the framework design. Other authors working on similar domains (e.g. [10] in the avionics domain) similarly stress the crucial importance of involving domain specialists in the development process.

The presumption of an involvement of domain experts in the framework development process leads to downplaying the "domain analysis" phase that is so often discussed in the framework literature. In a sense, domain analysis is assumed to have already been implicitly performed through the acquisition of domain expertise by the domain experts and this is assumed to have been done prior to and outside the framework development process. The problem facing the framework development team therefore lies in the decision as to which ones of the (known and well-understood) functionalities typically implemented by applications in their domain of interest should be modelled by the framework.

Later in the chapter, it is argued that the specification process for a framework should be iterative and should overlap with the framework design. To some extent, and due to the complexity of frameworks, their functionalities are not so much defined in advance as *discovered* during the design. It may therefore be unprofitable to put much effort into domain analysis and specification before the development starts since decisions taken at this time are likely to be overturned as a result of unanticipated findings during the design phase. An initial specification phase is still required to kick-start the design process but it should be understood and accepted that the exact contour of the framework will only become clear as the framework is developed.

Software frameworks are developed as reusable assets. They are intended to offer to application developers constructs that they can use to rapidly instantiate their applications. Obviously, the requirements of these individual applications are unlikely to be known at the time the framework is designed. A framework therefore is developed to serve future needs that can only approximately be anticipated. It is this uncertainty that surrounds the actual usage of the framework that militates in favour of a streamlined and non-overly formalized domain analysis and specification phase.

In chapter 7, it is argued that greater reuse will require a shift from a specification-driven to an explicitly reuse-driven approach. In a specification-driven approach, an application is developed to conform to a set of formal specifications that are tailored to a specific need and that were defined in advance of the application's development. In a reuse-driven approach, the customer who commissions the application, before specifying it, considers existing reusable products and attempts to tailor his specifications to them to facilitate their incorporation in his application. The latter approach is only possible if the reusable products are described in a manner that makes their evaluation for reuse easy. Since frameworks are meant to be reusable products, it follows that conceptual tools are required to characterize them from the point of view of their prospective users and to allow easy matching between them and their target application. There should exist, in other words, a formal description of the constructs offered by a framework. This implies a greater emphasis on *a posteriori* description of the framework which to some extent compensates for the diminished importance assigned to the *a priori* specification of the framework. It might be argued that *a posteriori* description and *a priori* specification are really one and the same thing. This is true in the sense that they contain the same information since both capture the externally visible properties of a product. Their target audiences, however, are very different. Specifications are aimed at the framework developers whereas the formalized *a posteriori* description of the framework is intended for its users, namely the application developers who might consider the framework as a starting point for the development of their application.

5.2 An Alternative Specification Approach

Specifying a framework is difficult because a framework is not something that – like a single application – can be run and tested. It instead represents a *set of potential applications* and it consists not only of concrete components but also of abstract constructs like interfaces and design patterns. A framework therefore must be described not only in terms of what it actually does but also in terms of what it might do if it were properly customized. Indeed, the second view is perhaps more important than the first because the essential feature of a framework is not its implementation of specific behaviour but its ability to be adapted to several different behaviours.

This means that the traditional approaches to software specification based on user requirements or use cases are not feasible since both user requirements and use cases are normally formulated (more or less explicitly) as outputs to be generated by a piece of software in response to certain inputs. This type of formulation is impossible with a framework which cannot be modelled as an input-output system. Finally, and more subtly, user requirements and use cases presuppose a development process driven by a specific need from which the requirements or the use cases arise. This is usually true in the case of applications but is unlikely to be true in the case of frameworks that tend to be developed as strategic assets for use in situations that evolve with time and that cannot be fully anticipated in advance.

If use cases cannot be used as such in the context of framework specification, the idea that underlies them – to specify an artifact by describing how the artifact is to be used – can indeed be transposed to software frameworks. Implementation cases as introduced in the previous chapter serve the same purpose for frameworks as use cases do for individual applications. Just as a sufficiently large number of use cases can be used as a specification of an application, a sufficient number of implementation cases can represent a specification for a framework. The only snag is that implementation cases are defined *during* the framework design process and it is therefore impossible to provide a set of implementation cases at specification time which – by definition – precedes the design of the framework.

Some means is therefore required to provide an initial input to the design process. Here we propose that this be done indirectly by specifying the target domain of the framework.

Figure 5.1 symbolically shows an application domain. The square boxes represent individual applications in the domain. The shaded boxes represent applications in the domain targeted by the framework, namely applications that the framework is intended to help develop. The framework target domain could in principle be fully described by listing all the applications in it. In practice this is normally not feasible. A more promising approach is to impose some structure upon the space of applications in the domain and then use a few application-points in this space to characterize the entire domain. The concepts of *ordering criteria* and *corner applications* in a domain are introduced for this purpose.

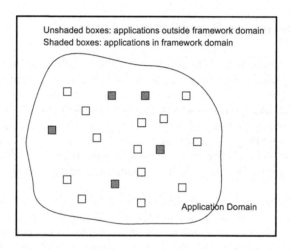

Fig. 5.1. Unstructured framework domain seen as a set of concrete applications of which some are in the framework target domain

Consider the set of all applications in an application domain. It will normally be possible to identify several criteria according to which the applications in this set can be ordered. For instance, one such criterion could be "degree of complexity" and the applications could then be ordered according to how complex they are. Another possible criterion could be "degree of distribution" and the applications could be ordered with at one extreme applications that are highly distributed and at the other extreme applications that run from a single node.

Figure 5.2 shows an application domain where two ordering criteria have been identified and are represented by the two cartesian axes. Each square box in the figure represents an individual application. The outer contour encloses all possible applications within the domain. The target domain for the framework is represented by the inner contour. It only encompasses a subset of the application domain as a framework will in general be targeted at only a subset of all possible applications in a certain domain.

The view taken here is that specifying the framework is analogous to specifying its target domain. Graphically, this is equivalent to defining the shaded area in the figure. If the space of applications is assumed to be a metric space – a space where the concept of distance can be defined – one way to define one of its subsets is through the *corner points* of the subset itself. For the example in the figure, the corner points for the framework domain are the applications A, B, C and D. The position of these four points defines the entire framework domain in the sense that it allows to decide which applications are within the set and which ones are outside it. Use of the corner points then is, in theory at least, an interesting candidate as a means of specifying the framework domain because it makes it possible to describe the entire domain using a finite set of concrete applications.

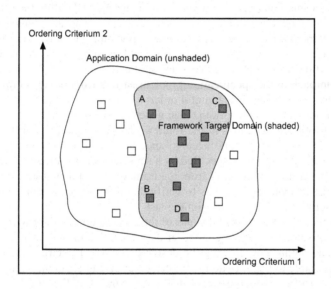

Fig. 5.2. Framework domain structured by ordering criteria

Obviously, use of the corner points to define the framework application domain only makes sense if the ordering criteria that impose a metrics on the space of all possible applications are "meaningful". Clearly, choosing non-meaningful ordering criteria (like, say, the number of comment lines in each application) leads to non-meaningful choices of corner points that cannot be used to define the target domain of the framework. Use of poorly chosen ordering criteria might result in a framework domain whose members – the concrete applications directly derivable from the framework – are randomly distributed over the application domain. Meaningful ordering criteria should result in a framework domain that looks compact and convex and that can therefore be economically described by a limited number of corner points.

The corner points of a framework application domain are important because their requirements can be used to specify the entire framework. If the ordering criteria in the domain space are well-chosen, then the requirements of the corner point applications can be taken as representative of the requirements of all applications that can potentially be generated by the framework. The underlying assumption is that if one framework can generate the corner points, then it can generate all the intermediate applications and can therefore generate any application in the framework domain. Use of the corner points thus replaces a potentially infinite set of requirements with a finite set (but this set could be large because it grows exponentially with the number of ordering criteria).

In summary then, this solution to the specification problem reduces the problem of specifying a framework to that of specifying the set of applications that can be generated from the framework – the framework target domain. This is done in two steps. First, a small number of ordering criteria are defined within the set of applications in the application domain and then corner points are identified that define the framework domain as a subset of the application domain. To these corner points there correspond concrete applications and the framework will meet its specifications if it can generate these concrete applications.

This approach to the specification problem suffers from obvious drawbacks: it is not always easy to build meaningful ordering criteria, it is impossible to provide a formal proof that a framework that can generate the corner applications can also generate all other applications in the framework domain, and it is difficult to verify that a delivered framework meets its specification. The last point refers to the fact that testing a framework specified by a set of corner point may require instantiating all the corner applications from the framework and then testing each one of them against its own requirements to verify that it is correctly implemented. This will in general be a very time-consuming task.

It must therefore be stressed that use of the corner point method is only suitable as a stopgap solution at the beginning of the framework development process and with the sole purpose of allowing the design process to be started. Afterwards, it becomes possible to begin to define implementation cases. As the design progresses, these must be continuously revisited and expanded by the stakeholders in the framework development process and constitute an evolving specification for the framework itself. At the end of the development process the worked-out implementation cases can be used to verify and demonstrate the correct implementation of the framework.

Figure 5.3 gives a schematic overview of the process proposed to specify a framework. As is evident from the figure, the specification of a framework becomes an iterative process. Iteration has two causes. On the one hand, it is a result of the high level of abstraction of frameworks. Specifying an artifact requires fully understanding the way it is to be used and such a full understanding is difficult to achieve when the artifact in question has a purely abstract nature. The way a framework is to be used gradually emerges as its design progresses. On the other hand, and as is often the case in software engineering, iteration is a consequence of complexity. This relationship is well known in the case of the design process for individual applications where it is normally accepted that, for all but the simplest applications, the software architecture has to be refined in successive steps. The specification process at application level is in theory supposed to be done in one single step. The user requirements should be an unchangeable input to the design. In practice, this is seldom the case and specifications are often tuned in response to difficulties encountered or opportunities discovered in the architectural design phase or even during software testing. Frameworks are considerably more complex than individual applications and it should not be surprising that this greater complexity entails a tighter coupling between the specification and architectural design phase. The use of implementation cases is the vehicle through which this coupling becomes explicit.

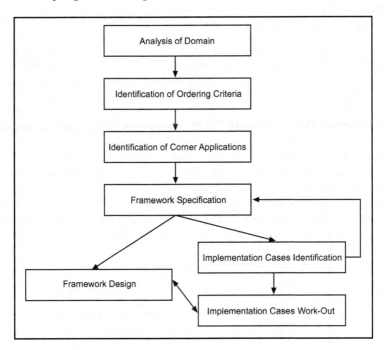

Fig. 5.3. Framework specification process

5.3 An Example from the AOCS Case Study

The domain of interest for the case study in the second part of the book is the set of all AOCS applications. When we first tried to delimit a target domain for the AOCS Framework, the first step was to clarify what is meant by "AOCS application". Satellite on-board software – like most embedded control software – is usually given a hierarchical structure where the three major software layers shown in figure 5.4 can be recognized [96]. Reusable solutions already exist for the bottom two layers – the device drivers and the operating system – and we therefore decided to concentrate on the top layer only and the term "AOCS Application" should accordingly be understood as designating the control software exclusive of the operating system and the device drivers. Within the space of all AOCS applications thus understood, we then had to identify a subset to serve as starting point for the design process. This initial specification of the domain was then refined as the design progressed. The method of ordering criteria and corner applications was used. In the case of the AOCS Framework, the following ordering criteria were adopted:

– Degree of on-board autonomy of the AOCS
– Mission complexity
– Centralized or decentralized architecture

The first criteria refers to the autonomy or ability of the satellite to operate for a specified period of time outside ground contact. The length of this period is often a driver for the on-board software and it can therefore be a good ordering criteria. In analyzing the second criteria, the mission purpose was used as a proxy for mission complexity. Typical mission purposes in rough order of complexity are "telecommunication" (least complex), "earth observation", "scientific observatory", "interplanetary mission", "technology demonstrator". Finally, the last criteria refers to the fact that there are two fundamental architectures for AOCS systems: *centralized* architectures where the AOCS software resides on the same processor as the central satellite software and *decentralized* architectures where the AOCS software resides on a dedicated processor. Intermediate solutions are also possible (though rare) as in the case where the two software are separate but do not have a specific processor (i.e. the processor on which they run is selected dynamically).

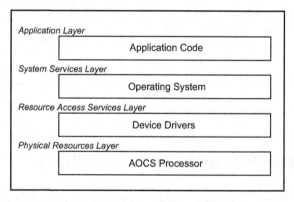

Fig. 5.4. Typical structure of the software for satellite control systems

Given these criteria, one could place all AOCS applications in a 3-dimensional space where autonomy, complexity and architectural centralization vary on each axis. At the beginning of the framework design process, we went through the exercise of building such a space using a number of recent satellite missions as reference points and we then selected from among them the "corner applications" for the framework. The resulting target space for the framework can be roughly described as follows:

- decentralized architecture
- autonomy level up to the level of the AOCS of the ISO or Rosetta satellites[11]
- complexity level up to the level of interplanetary missions

Description of the corner missions that were selected to characterize the above domain goes beyond the scope of this book but it is important to stress that analysis of their user requirements (the AOCS software is usually specified by means of user requirements) was the first input to the framework design process. After the design started, implementation cases were gradually created and elaborated and became the main drivers of the framework specification process.

---

[11] ISO and Rosetta are two scientific missions developed by the European Space Agency. They were among those we considered when analyzing the AOCS domain.

6 Framework Design

This chapter proposes a multi-stage design procedure for software frameworks that is based on the use of the framelet and implementation case concepts presented in an earlier chapter.

6.1 Overall Approach

Design definition covers the phases in the software development process that lie between specification and coding. Traditionally, the gap between them is filled by one single conceptual step usually called architectural design. Its objective is to identify the modules or the classes that can best encapsulate the behaviours to be implemented by the target application. Sometimes the architectural design phase is divided into subphases, perhaps looking at individual subsystems, but their objectives and tasks are essentially the same as those of the overall architectural design phase although at a different level of granularity.

In a first innovation, we propose to introduce one additional phase called *concept definition phase* that precedes the architectural design phase. The introduction of a concept definition phase as distinct from the architecture definition phase reflects the distinction between design and architecture already made in chapter 3. Its objective is to define the framework at design level. The design solutions that are offered in this phase are characterized by at least three features.

- The abstract interfaces and classes that enter in the solutions are defined in an incomplete manner. Only key methods are described.
- The system is broken up into design units (the framelets) and solutions for their design problems are defined for each unit independently of the others. No effort is made to integrate solutions to different design problems.
- The main thrust of the design effort is on finding design patterns. The search for solutions at the concept level is essentially a search for design patterns to model the hot-spots identified during the domain analysis. This search may start from standard patterns but it will normally progress to the definition of new patterns that are optimized for the particular domain of interest.

When a framelet approach is adopted, the concept definition phase must be split into two phases: framework concept definition and framelet concept definition.

A. Pasetti: Software Frameworks, LNCS 2231, pp. 79-98, 2002.
© Springer-Verlag Berlin Heidelberg 2002

Some of the steps assigned to the concept design phase are probably taken by most experienced designers as a matter of course when they design complex individual applications. They are thus not unique to framework design. It is, however, useful to recognize them as formal phases in the framework design process because this allows them to serve as project milestones. This means that concept level decisions are to be documented and subjected to formal review by fellow designers and by customers. Given the complexity of the framework design process, this extra design and review milestones will be very beneficial.

It should also be noted that the effort required to arrive at a finalized architectural design as a percentage of the total project effort is much larger in the case of frameworks than in the case of individual applications. The "investment" made in an architectural design is therefore much larger in the case of frameworks than in that of individual applications. In order to protect this investment and to ensure that the effort is well-spent, it is wise to have one extra review layer before arriving at the architectural design milestone.

In conventional methodologies, the architectural definition phase is usually seen as separate from the specification phase. In a second innovation, and as already seen in the previous chapter, we see the specification of the functionalities to be offered by a framework as a continuous process that extends into the concept and architectural design phases.

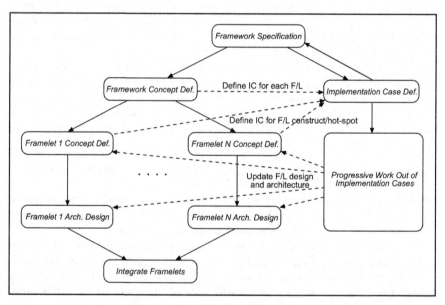

Fig. 6.1. Activity flow for the framework design process (IC = implementation case; F/L=Framelet)

Figure 6.1 shows the proposed framework design process. The main design tasks are enclosed in rounded boxes. Solid arrows represent temporal transitions from a design task to another and dashed arrows represent the main interactions among different design tasks. After the framework is specified, the design splits in two branches that are executed in parallel. In the left-hand branch, the framelets are designed and then integrated to form the framework. The activities in this branch are described in greater detail in this chapter. In the right-hand branch, the implementation cases are first defined and then gradually worked out. The work-out task is enclosed in an elongated box to explicitly show that it is a gradual process that takes place over a prolonged period of time spanning the entire framelet design phase. Note also how there is a feedback link from the implementation case definition box to the framework specification box. This represents symbolically the use of implementation cases as a specification vehicle for the framework.

The left-hand branch in the figure is in turn divided into many sub-branches, one for each framelet. Framelet design is articulated over two phases: concept definition phase and architectural design phase. In order not to clutter the figure, only a unidirectional flow of control work is shown but it should be clear that a software design procedure is inherently iterative and design refinement will usually require the procedure to be repeated either in part or in full.

The procedure of the figure covers only those aspects of the design process that are specific to frameworks. As the design proceeds, the framework comes to resemble a system of interacting classes of the kind that are commonly encountered when developing single applications. A dedicated framework design procedure then becomes redundant and developers should gradually turn to one of the many conventional methodologies that have been elaborated to guide the development of ordinary applications. The weakest point of the procedure we propose lies perhaps precisely at the junction between old and new methodologies. The difficulties in the transition between the two are symbolized by the difficulty of transforming framelets into subsystems. Framelets simplify the design process by breaking up the framework into smaller units of design represented by sets of related abstract interfaces and design patterns. Framelets only exist at the design level and will generally lose their identity – and much of their usefulness – as the framework moves into the architectural definition phase where simplification is achieved by dividing the framework into subsystems that gather together groups of interacting classes and/or components. The *conceptual* border between framelets and subsystems is clear but the practical implications of its existence less so. More experience will be needed before some general concrete guidelines can be derived on how to make the transition from the design to the architectural level and the corresponding hand-over from a framework-oriented to a conventional methodology. A more concrete discussion of this problem in the context of the AOCS Framework case study can be found in sections 8.9 and 8.10.

6.2 Alternative Approaches

In order to throw into relief the specificities of the design procedure of figure 6.1, it is useful to compare it with alternative approaches proposed by other authors. Perhaps its chief distinctive feature lies in the explicit differentiation between architecture and design. To some extent, the activities prescribed for the concept definition phase – the phase where the framework design is defined – are also recommended by other authors. Reference [29], for instance, with the so-called FAST approach recommends that framework development should be preceded by an analysis of variability and commonality in the framework domain. Other authors (e.g. [63, 31]) make similar recommendations. The concept phase, however, goes further because in addition to identifying the commonalities and the points of variability, it must propose domain-specific design patterns to encapsulate them.

Virtually all framework researchers stress the cyclical nature of framework development. One could argue that what is called here the design or concept stage is no more than what others see as the first iteration in the architectural definition phase. In principle, this view is acceptable (unless, of course, one has a penchant for semantic disputes). Nonetheless, explicit identification of a design stage is regarded as worthwhile because it draws attention to the multi-stage character of the process through which software architecture is constructed. When dealing with single-product development, this character is not strongly in evidence because the various iterations in the architectural definition phase are conceptually similar: they are all aimed at identifying objects and their relationships and successive iterations differ merely in the degree of granularity of the description they offer. In the case of frameworks, however, the development process clearly divides into separate phases. The task of identifying the hot-spots and their matching design patterns is conceptually very different from that of defining object architectures, whence the value of giving different names to the associated development stages.

One consequence of the introduction of a concept phase devoted to attacking design problems is that design simplification is achieved by introducing framelets as self-standing units of design and by breaking up the process of designing the framework into parallel and simpler framelet design processes. Other approaches, by contrast, more or less tacitly assume that design simplification is to be achieved by considering clusters of related classes or components.

To some extent, lack of emphasis on the design phase as such could be compensated with a description of the architecture that is more comprehensive than that proposed here. Reference [31] for instance advocates the use of multiple "architectural views" (in the sense of reference [9]) as a way of mastering framework complexity both from the perspective of the designer and from that of the user. Reference [30] takes a similar approach with each view addressing a specific set of concerns of a system.

A different approach to subdividing the framework design process into simpler and more manageable units is based on aspects. Reference [58] proposes to use aspects to identify quality attributes (such as performance or reliability), isolate the requirements that pertain to a specific aspect and develop an architectural style to address it. In a second phase, the various architectural styles are merged together. Given the relationships between framelets and aspects (already noted in section 4.1.3) and the similarities between architectural styles and design patterns, the approach of this author is very close to that proposed here.

The central place of design patterns is a second distinctive feature of the design strategy proposed here. Other authors who assign an explicit role to design patterns tend to see them either as ways of documenting fundamental structural design models as in [29] or consider only generic (as opposed to domain-specific) design patterns and restrict their usefulness to the encapsulation of best practice [93]. One work that gives them a prominent role is [20]. Its author proposes that the design process should start with a first coarse-grained definition of the system architecture essentially based on the identification of so-called *archetypes*. Archetypes capture key domain abstractions but, although they usually do not correspond to directly identifiable entities in the target applications, they are encapsulated in objects. The architecture thus obtained is then subjected to a number of transformations aimed at improving its quality. Two transformations that are relevant in the present context are: the imposition of architectural patterns and the application of design patterns. The two types of patterns differ in their scope. Architectural patterns are like design rules that are applied to the entire system and that address specific system-wide problems. Design patterns are instead applied locally. Our design procedure conflates the two concepts because it explicitly regards them as a way of enforcing large-scale architectural uniformity by ensuring that similar problems in different locations in the domain receive similar solutions. The main difference between its approach to the design problem and that of reference [20] is that the latter starts from an object-based architectural draft and uses the patterns – of both the architectural and design type – to transform it whereas we start from the design pattern and derive the object representation of the architecture from them. Furthermore, we try to split the design problem into simpler units whereas reference [20] takes a more holistic view. This is the reason why the concept of architectural pattern, in the sense of a design rule in-the-large that covers the entire system, does not really exist in the design procedure we propose (at least not at design level during the concept definition phases).

Finally, we articulate the design process upon two parallel threads. On the one side, the framework design proper is performed while, on the other side, implementation cases are worked out as a means of checking the adequacy of the design. This emphasis on continuous design monitoring is a further distinctive feature and strength of the procedure we propose.

6.3 The Framework Concept Definition Phase

This is the first phase in the framework design process of figure 6.1. Its primary input is the initial framework specification as obtained from an analysis of the framework domain, perhaps using the method of the corner applications or some other way of roughly demarcating the target domain for the framework. Its broad objective is to define a design concept for the framework.

The concept definition phase can be divided into subphases giving rise to the flow of activities shown in figure 6.2. The figure shows that there are three concurrent threads to the framework concept definition phase. In one thread, the general design principles are defined. In a second thread the domain abstractions and the domain model are developed. This is done in parallel to the identification of the hot-spots, design patterns, framelets and implementation cases which are done in the third thread. Obviously, there is much conceptual interaction between the second and third threads. All threads are concluded with an investigation of design alternatives. The individual activities (the boxes in the figure) and their outputs are described in the following subsections.

The output of the concept definition phase is a *framework conceptual definition document* which describes the framework at design level. Its specific outputs are as listed in the subsections below. The second part of this book gives an overview of the AOCS Framework at a level of detail that is intermediate between framework concept definition and framelet concept definition.

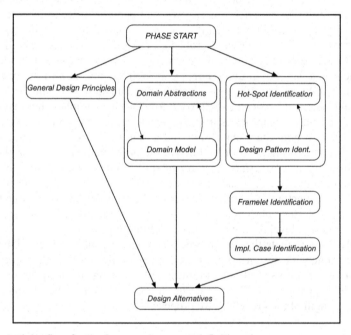

Fig. 6.2. Activity flow for the framework concept definition phase

6.3.1 Definition of General Design Principles

The general design principles are the overall constraints on the framework design. They define the policy to be adopted in regard of general issues like: use of multiple inheritance, use of dynamic memory and task allocation, use of non-standard language constructs (e.g. run-time type information), use of templates, adaptation mechanisms (e.g. composition or inheritance), general class structure (e.g. should a root object be defined? Should multiple inheritance be forbidden?), allowed implementation languages, use of component standards, etc.

This step is especially important in the case of real-time or other critical application domains where the framework designers may need to restrict the type of architectural constructs to be used in the framework development in order to increase the predictability or safety of end-applications.

The output of this step is a list of the proposed design principles together with a justification for their selection. A summary of this output for the AOCS Framework case study can be found in chapter 9.

6.3.2 Identification of Domain Abstractions

In this subphase, a *domain dictionary* is built up to describe the applications in the framework domain. This is done by identifying and defining the key terms that are found to recur in the description of several individual applications. These key terms correspond to objects and concepts that exist in most target applications and which therefore have to be modelled by the framework. The importance of building up a domain dictionary early in the development process is stressed by many authors [29, 63, 47, 82, 2]. The preparation of the domain dictionary is crucial not only as an implicit step in the modelling of the framework domain but also to provide a common vocabulary to allow designers and other framework stakeholders to communicate effectively and efficiently.

Note the difference between the domain dictionary defined here and the data dictionaries required by many application design procedures. The latter gather concrete concepts used in concrete applications. The former defines abstractions that span applications. For instance, in the case of the AOCS Framework, the manoeuvre concept was defined as an orderly sequence of actions to be performed by the AOCS. This concept only exists at the framework level since individual AOCS applications define only *concrete* manoeuvres (e.g. an attitude slew, a wheel unloading, etc) and do not need the abstract manoeuvre concept.

The output of this step is a *domain dictionary* listing the framework abstractions and their definitions. The domain dictionary for the AOCS Framework is given in the appendix.

6.3.3 Construction of the Framework Domain Model

This activity is in practice conflated with the previous one since in order to identify the domain abstractions it is necessary to have a model of the application domain and, conversely, a model of the application domain can only be described in terms of domain abstractions. Indeed, in practice, one good way to build a domain model is to analyse the entries in the domain dictionary [55].

Construction of the framework domain should not be confused with the specification problem of the previous chapter. The domain model is built on the assumption that the domain has already been delimited during the specification phase (perhaps using the corner applications or by defining use cases). Its objective is to organize the information provided by the specification phase in a manner that is useful for the framework design and development. Like the domain dictionary, it serves primarily as a common basis for discussions within the development team and with the framework customers.

The output of this subphase is a description in informal language of the framework domain. An abbreviated version of this output for the AOCS Framework is given in chapter 2.

6.3.4 Identification of Framework Hot-Spots

Individual applications are obtained from the framework by customizing its hotspots. The aim of this activity is to identify the points where this type of behaviour adaptability has to be built into the framework. The key question that must be answered is: where in the framework should behaviour tuning be available in order to enable the framework to generate all the applications in the framework domain?

Hot-spot cards [80] are one practical mechanism to answer this question and identify the hot-spots. They are a means to let framework designers elicit information from domain specialists about typical variability patterns in the applications in the framework domain. Their function is similar to that of CRC cards but at a higher level of abstractions: CRCs are intended to describe behaviours associated to individual classes; hot-spot cards are intended to describe a system of cooperating classes that parameterize behaviour and allows behaviour tuning.

The output of this subphase is a list of hot-spot. A complete and cross-referenced list of the hot-spots of the AOCS Framework is found in the appendix.

6.3.5 Identification of Framework Design Patterns

In the framework conceptualization adopted here, the definition of domain-specific design patterns has been recognized as crucial to the design of a good framework. In this subphase, a catalogue of generic design patterns is first identified (such as the classic references [44, 24]) and it is then scanned to isolate patterns that could be applicable to the framework. Inadequacies in the standard patterns from the selected catalogues and the need for more domain specific design patterns are also identified. In general, at this stage, a design pattern should be identified for each hot-spot.

The output of this subphase is a list of the selected patterns, of their relationship to the hot-spots and of how they can be applied to the framework. The design patterns developed for the AOCS Framework are presented later in the book with a complete and cross-referenced list being available in the appendix.

6.3.6 Framelet Identification

In this subphase, the framework is broken up into individual framelets. The heuristics proposed in section 4.1.6 can help this task. The function of each framelet and their mutual relationships are also defined.

The output of this subphase is an overview of each framelet and of its interactions with other framelets. For the AOCS Framework, this output is provided in the case study part of the book.

6.3.7 Identification of Implementation Cases

For each framelet, the basic implementation cases are identified. Eventually, implementation cases should cover all the functionalities offered by the framework. At this stage, they should cover all design patterns and all hot-spots identified at framework level. The degree of the definition of the implementation cases can only be sketchy and initially can be restricted to a bare statement of their objectives.

The output of this subphase is a list of implementation cases. Section 4.2.3 proposes a template of how implementation cases can be described. Some implementation cases are discussed in the context of the AOCS case study.

6.3.8 Identification of Alternative Solutions

Framework design is necessarily an iterative process. In order to facilitate the iterations, it is necessary to systematically list design alternatives at every step of the design process. Thus, each of the outputs required for this phase is accompanied by a description of its alternatives. Discussion of selected design alternatives for the AOCS Framework case study is presented in the second part of the book.

6.4 Framelet Concept Definition

This is the second phase in the framework design process (see figure 6.1). It is performed for each framelet in the framework. Its objective is to define the framelets at design level. This means that design solutions must be found for the problems defined by each framelet. At this stage, however, the framelets are largely considered in isolation from each other and each design problem and its proposed design solutions are fully self-contained. The flow of activities in the framework concept definition phase and their temporal sequence are shown in figure 6.3. As in the case of the framework concept definition, three parallel activity flows are recognized. The first one covers the identification of the abstract interfaces and interface implementations exported by the framelet. The second one covers the identification of the framelet hot-spots and the definition of the design patterns.

These two activities are closely coupled because design patterns are often associated to hot-spots and viceversa. Finally, in the third flow of activities the relationship of the framelet to the framework is defined.

The activities shown in the figure are described in the subsections that follow. In reading them, the difference between *identification* of a feature and its *definition* should be noted. Identification requires simply that the presence and need for the feature be recognized and that its function be informally and perhaps incompletely described. Definition instead implies full formal description of the feature. In general, features are first identified and then defined as the design matures. Because of their higher level of abstraction, design patterns should be defined in this phase. Other features – notably hot-spots, abstract interfaces and interface implementations – are only identified and are defined in the architectural design phase.

The output of this phase is, for each framelet, a *framelet concept definition document* that addresses the points listed in the subsections below.

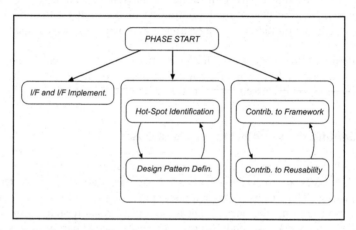

Fig. 6.3. Activity flow for the framelet concept definition phase

6.4.1 Identification of Exported Interfaces and Implementations

Framelets export three types of constructs: abstract interfaces, design pattern and interface implementations encapsulated within components (see section 4.1.5). In this phase, the abstract interfaces and interface implementations exported by a framelet are identified.

6.4.2 Identification of Framelet Hot-Spots

As discussed in section 4.1, framelets partition the space of hot-spots and it is therefore possible to assign the framework hot-spots to the framelets. In this activity, the hot-spots where framelet behaviour can be adapted to the needs of their environment are identified. For each hot-spot, an adaptation mechanism and an adaptation design pattern are proposed. Note that while all framework-level hot-spots must map to some framelet-level hot-spot, the converse is not true. Framelets offer hot-spots that function as handles for the integration of other framelets. Such hot-spots are not present at framework-level.

6.4.3 Definition of Applicable Design Patterns

Design patterns are the third type of architectural construct exported by the framelet. In this phase, the design patterns that can be used from existing catalogues are identified. Standard catalogue patterns aim at generality of applicability. Often, it will be possible to particularize them to the needs of the framework target domain. When no suitable standard pattern can be found, new design patterns must be identified. At the very least, a design pattern has to be proposed for each of the major hot-spots identified at the framework concept level.

This is one of the crucial steps in the framework development. Design patterns are seen as the building blocks of the framework one of whose functions is to offer a repository of domain-specific design patterns that are tailored to the needs of the framework domain. Standard catalogue patterns, precisely because they are not aimed at any specific domain, tend to offer what might be called "high-level design solutions" which are often accompanied by implementation options and by design alternatives. This activity largely consists in refining these standard patterns to transform them into domain-specific patterns which should ideally be "ready to use" in the sense of leaving as few implementation or design alternatives open as possible. This concept is further illustrated in the context of the AOCS case study in section 8.8.

6.4.4 Definition of Framelet Contribution to the Framework

In the architectural definition phase, the framelets will be integrated together to form the full framework. In this subphase, their mutual relationships and their functions within the framework are reviewed. Its objective is a preliminary check that integration is feasible and that there are no fundamental incompatibilities between the design solutions proposed by different framelets.

6.4.5 Definition of Framelet Contribution to Reusability

The purpose of a framework is to promote software reusability. In this activity, the way in which each framelet contributes to this goal is explicitly described. This helps identify redundant framelets or framelets that need to be improved. This exercise may seem unnecessary but important design decisions – such as the decision to introduce a framelet – have a way of crystallizing and need to be periodically challenged to verify that the reasons that prompted them are still valid. In the case of the AOCS Framework, for instance, forcing the designers to spell out the contribution of each framelet to reusability eventually led to the removal of two framelets as no longer necessary.

6.5 Framelet Architectural Definition

The objective of the previous two phases was to define the framework *design*. The objective of this phase is to define its *architecture*. Its primary conceptual input are the design solutions – usually in the form of design patterns – identified for the framework. Its task is to instantiate these design patterns and to fully define all the classes and interfaces required to support them. In the concept definition phase, design problems were addressed in isolation from each other. In the architectural phase, these solutions must be combined and integrated.

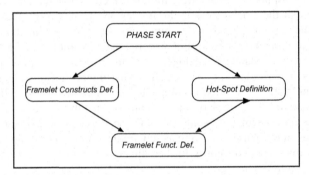

Fig. 6.4. Activity flow for the framelet architecture definition phase

The flow of activities of the architectural design phase are shown in figure 6.4 in their temporal sequence. These activities are performed separately for each framelet and result in a phase output of, for each framelet, a *framelet architectural definition document*. In practice, since the transition from the concept to the architectural definition phase is incremental and gradual, it is recommended to expand the framelet concept definition document that in the end will document both the concept and the architecture of a framelet. This is the approach that was followed for the AOCS architecture. One drawback of this approach is that, since at architectural level the framelets have been merged, the framelet architectural definition documents are no longer independent of each other. The phase activities are briefly described in the following subsections.

6.5.1 Definition of Framelet Constructs

The constructs exported by the framelet – abstract interfaces, design patterns and interface implementations – are identified in the concept definition phase. In that phase, a full definition is only given for the design patterns. In this phase, the abstract interfaces and the interface implementations are defined and the design patterns are instantiated. The difference in treatment between design patterns on the one side and interfaces and components on the other is due to their different levels of abstraction. Design patterns pertain to a higher level of abstraction than interfaces and components while their *instantiation* is coterminous with the definition of the abstract interfaces and components that embody the design pattern in a concrete architecture.

Definition of interfaces and components is done at the level that is typical of the architectural definition phase in conventional methodologies. Broadly speaking, this means that the semantics and signatures of the operations offered by the abstract interfaces is described and that the external interfaces of the components encapsulating the interface implementations are described.

The full definition of the framework classes implies the integration of the framelets. Framelets are integrated when concrete components are defined that implement several abstract interfaces belonging to different framelets and that participate to several design patterns belonging to different framelets. For an example from the AOCS Framework, consider the controller manager component. The need for this component emerges during the elaboration of a design solution to the controller management problem in the controller framelet (see section 8.4 or chapter 22). During the concept definition phases, its key methods and functionalities are described but no attempt is made to define all its methods and its relationship to other framework components. This is done at architectural level where, among other things, it is decided that the controller manager should implement interface Telemeterable which is introduced by the telemetry management framelet (see section 8.3). At this point, the component straddles two framelets – the controller framelet where it was first defined and the telemetry framelet that exports the Telemeterable interface – and thus becomes an agent of integration between them (see also the discussion in section 8.10).

6.5.2 Definition of Framelet Hot-Spots

The framelet hot-spots were identified in the concept definition phase. They are now formally defined according to the guidelines discussed later in the chapter. Essentially, this means that the hot-spots are associated to specific architectural constructs (classes that must be subclassed, flags that must be set, plug-in components that must be loaded, etc.) and that the adaptation mechanism is defined.

6.5.3 Definition of Framelet Functionalities

The concept of functionality is introduced in the next chapter. Functionalities are proposed as a means through which a framework is described from the point of view of a prospective user. They are designed to allow application developers to quickly and efficiently match the requirements of their application to the offerings of the framework with the purpose of facilitating their decision as to whether the framework could be used to aid the development of the application.

The functionality concept is rather tentative and other mechanisms might be considered in their stead. The point to stress here is that one of the task of the architectural definition phase should be a formal description of the constructs offered by the framework specifically targeted at application developers.

6.6 Framework Design Description

A design description formalism for frameworks has to recognize that the design of a framework goes through several formal phases each with its own distinct modelling needs. Accordingly, the next three subsections discuss the design description problem for the three design phases identified in the first part of this chapter. The description approach is summarized in figure 6.5.

6.6.1 Framework Concept Definition

In the framework concept definition phase, the framework architecture and the design decisions exist at a high level of abstraction where plain, informal English is the best modelling option. Our experience is that the design at this stage is so unstable that any attempt to formalize it is counterproductive: there is no point in formalizing something that is still vague and subject to continuous change.

Formal design modelling is moreover aimed at describing architectural constructs that have been *defined* but in this stage the emphasis is on *identifying* architectural constructs, namely in recognizing the need for them and in searching for analogous problems in other domains that might provide tips for solutions. Definition will be done in later phases of the design process. The framework design process can be seen as a gradual evolution from the identification of abstract constructs – mainly design patterns and interfaces – towards their definition and embodiment in concrete components and fully specified interfaces. The identification phase is not formally recognized in conventional methodologies targeted at application development and, probably for this reason, design modelling techniques are not good at handling concept identification. At any rate, it is questionable whether any kind of formalism would be useful at the beginning of the design process since, at this stage, *notions* should be more important than *notation* and notions are often best expressed in plain English.

6.6.2 Framelet Concept Definition

In the framelet concept definition phase, some degree of formalism and systematic treatment becomes necessary at least for three of the five tasks applicable to this phase (see figure 6.3), namely the identification of exported abstract interfaces and interface implementations, the definition of design patterns, and the identification of hot-spots.

Abstract interfaces and components can be described with any of several available formalisms of which the most popular is UML. It should be noted that UML does not predefine any formalism for describing framework constructs. Some authors have proposed using the UML meta-language extension mechanism to develop such a formalism [41]. This may be adequate at the architectural design phase but is unnecessary at this phase where ordinary UML diagrams are sufficient to convey the framelet concepts.

Description of the framelet design patterns is problematic since no accepted formalism exists for design pattern modeling. In some cases, the design patterns will come from a pattern catalogue in which case reference can be made to the description in the catalogue. When new patterns are instead used, they can be described in informal language, perhaps following the model of [44] that has becomes a *de facto* standard for design pattern description.

Finally, the framelet hot-spots must be systematically classified. The classification scheme we propose is the one adopted in the AOCS Framework project and extends the hot-spot card scheme of [80] by adapting it to the needs of frameworks as conceptualized here. More specifically, for each hot-spot, the following information must be provided:

- *The visibility level*: two visibility levels are possible: *framelet-level* or *framework level*. Some hot-spots exist only at the framelet level as they are intended to provide hooks for other framelets during the framework assembly process. Such hot-spots are said to have a framelet-level visibility. Other hot-spots carry over to the framework level as they are intended as hooks where application developers can insert application-specific items during the application instantiation process. Such hot-spots are said to have a framework-level visibility.
- *The adaptation time*: hot-spots provide a means of adapting framelet behaviour. Two adaptation times are possible *compile-time* and *run-time*, depending on whether behaviour adaptation is done statically (e.g. using inheritance or template instantiation) or dynamically (e.g. using composition).
- *The adaptation method*: a hot-spot is a point where framelet behaviour can be adapted. The following adaptation mechanisms (loosely based on that of reference [42]) can be identified:
  - enabling/disabling a feature (component configuration)
  - tuning an existing feature (component configuration)
  - defining a functionality implementation (component plug-in/subclassing)
  - replacing a feature (method override)
  - augmenting a feature (method override/subclassing)
  - adding a new feature (subclassing)
- *The pre-defined options*: in some cases, the framelet itself offers pre-defined options for a hot-spot. For instance, the control algorithm in an attitude controller component is clearly a hot-spot because different applications have different types of control algorithms. However, the framework may offer some plug-in components implementing common types of control algorithms. These

default offerings are the default components identified in section 3.2 as one of the three types of basic constructs that make up a framework.

Reference [29], in discussing how the points of variation in a framework domain should be characterized, stresses the need to provide bounds to the variability in order to constrain the flexibility of the framework and hence simplify its design. This type of information makes sense in the case of parameter-like hot-spots, namely hot-spots where adaptation consists in configuring a parameter by setting its value. However, our experience is that most major hot-spots (and in the concept definition phase only major hot-spots are identified and classified) tend to be feature-like rather than parameter-like in the sense that they are described by plug-in components or by methods to be overridden. The concepts of "variability bounds" is probably not very useful for this type of hot-spots. In any case, variability bounds come into play only when developing default options for the hot-spots (like default components or default implementations for abstract operations) and this is done in the architectural phase.

Reference [90] adds one further dimension to hot-spot categorization modelled on the meta-pattern concept [76]. Its author recognizes three types of hot-spots. *Not-recursive* hot-spots in which a requested service is provided by only one component; *chain-structured (1:1) recursive* hot-spots in which a requested service may be provided by a chain of sub-class components in the style of the chain of responsibility pattern; and *tree-structured (1:n) recursive* hot-spots where the requested service may be provided by a tree of sub-class components in the style of the composite pattern. We have no direct experience of using this categorization but it seems both informative and easy to handle and deserves to be taken into consideration.

6.6.3 Framelet Architectural Definition

In the framelet architectural definition phase, a full formal definition of the framelet features is necessary. Reference [41] introduces a UML-based formalism (UML-F) for this purpose that is tailored to the need of frameworks. Its chief strength lies in the UML augmentation that it defines to characterize the points of adaptation in the framework classes and to map design patterns to the framework classes. This provides a unified approach to the problem of framework description at architectural level. The same UML diagrams can be used to describe all aspects of the frameworks: the abstract interfaces, the components and design patterns it exports and the hot-spots that it exposes to the application developers.

6.6.4 Overview of Design Description Techniques

Figure 6.5 summarizes the design description process we propose. Solid arrows represent transitions from one modelling or design phase to another. Dashed arrows represent information interactions between design and modelling phases.

The design starts with the system concept definition to which there corresponds a design description in informal language. The next design phase is the framelet concept definition to which there correspond three modelling tasks to describe, respectively, the framelet design patterns, its components and interfaces, and its hot-spots. In this phase, design patterns and hot-spots are described informally but systematically. Component and interfaces are described using standard UML. Finally, the framelet architectural design is described using extended UML diagrams such as those prescribed by UML-F or other UML extensions powerful enough to capture the semantics of design patterns.

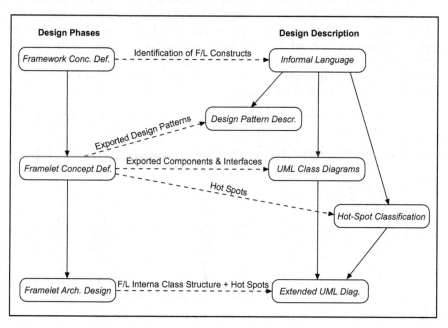

Fig. 6.5. Framework modelling as a function of the design phase (F/L=Framelet)

6.6.5 Framelet Interactions

We have broken up framework design into the design of the individual framelets. Framework modelling must accordingly be reduced to the problem of modelling the framelets. However, at some point, the framelets must be combined together and therefore framework modelling should include a description of the mutual interactions of the framelets.

From an external point of view, the framelets are characterized by the constructs they export. For modelling purposes, it is useful to distinguish between "horizontal" and "vertical" export of constructs. Constructs are exported horizontally to other framelets and vertically to the applications that are instantiated from the framework. To illustrate, consider figure 6.6 which shows three framelets from the AOCS Framework together with some of the constructs they export. The mode management framelet exports one design pattern to the telemetry framelet. This means that in defining the internal architecture use is made of the mode management design pattern or, in other words, this design pattern is instantiated within the telemetry framelet. The telemetry framelet exports constructs both horizontally and vertically. The telemetry manager component and the telemeterable interface for instance are exported vertically by which it is meant that developers of AOCS applications are expected to make use both of the telemeterable interface and the telemetry manager component in assembling their application. Note incidentally that the telemeterable interface is exported both vertically and horizontally. The interactions of framelets could therefore be described by diagrams like those of figure 6.6. Although we did not investigate this matter, is seems likely that the UML meta language mechanism could be used to introduce a formalism for describing this kind of diagrams thus allowing framelet interactions to be described within UML using standard UML tools.

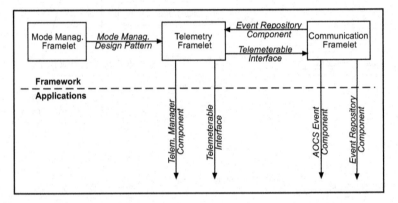

Fig. 6.6. Example of framelet interactions through the constructs they export

6.6.6 Examples from AOCS Case Study

At the end of the design process, for each framelet, there should be a *framelet concept and architectural description document* that provides all the information described in this chapter. The AOCS Framework web site gathers together such documents as they were produced in the AOCS Framework Project. Their content is standardized and always includes the following items:

- a table listing the constructs exported by each framelet
- UML class diagrams describing the class structure of the framelet
- for each framelet interface, a table describing the semantics of its methods
- for each hot-spot, a table describing the hot-spots

The UML class diagrams and method descriptions are of course well-known. For an example of a framelet construct table consider table 9 taken from one of the

AOCS framelet description documents. All framelet design documents should begin with a table like this one listing all the constructs exported by the framelet with a brief description. The constructs are divided into four categories. Each construct is then described in detail in the body of the document. When several large and related frameworks are used, databases can be set up to act as archives for the frameworks. Search engines could be added to allow quick location of constructs.

In the case study part of the book – at the end of each of the chapters covering the framelets of the AOCS Framework – preliminary versions of the framelet table of constructs are presented. They are obviously incomplete because they are based on the state of the framework as it exists during the concept definition phase but they can be a starting point for the definition of the full tables as shown in Table 9.

Table 9. Example of framelet construct table

OBJECT MONITORING FRAMELET
Design Patterns
Property Definition Pattern : pattern to define properties in objects and the methods to access them
Additional Properties Pattern : pattern to add new properties to a component that is already packaged as a binary unit (not used in prototype framework)
Direct Monitoring Pattern : pattern to directly monitor an object's property
Monitoring through Change Notification Pattern : pattern to implement a notification mechanism when a property changes in a specified manner.
Framelet Interfaces and Abstract Base Classes
`ChangeObject` : abstract base class for objects encapsulating a type of property change
Framelet Core Components
`Property` : encapsulation of a property
Framelet Default Components
`SimpleChange` : implementation of interface `ChangeObject` encapsulating a simple change in a property value
`OutOfRangeChange` : implementation of interface `ChangeObject` encapsulating an out-of-range change in a property value
`DeltaChange` : implementation of interface `ChangeObject` encapsulating a delta change in a property value
`SpikeFilteredDeltaChange` : implementation of interface `ChangeObject` encapsulating a delta change in a property value with spike filtering (not implemented in prototype framework)

This table is taken from the concept and architectural description document of the object monitoring framelet in the AOCS Framework.

Table 10 gives an example of a description of one of the hot-spots of the AOCS Framework. The table is taken from the concept and architectural description document for the intercomponent communication framelet. Each framelet document should carry a list of tables like this one describing the hot-spots in the framelet. In the AOCS Framework, a typical framelet has between 5 and 10 hotspots. After a framework has been fully instantiated, all its hot-spots should have been filled in. Projects could therefore set up and maintain hotspot databases that record both the hot-spot characteristics as shown in the table and the way the hotspot is being filled in for a particular project. Project completion can then be chekked by verifying that all hot-spots have been filled.

Table 10. Example of hot-spot description

Name: **AOCS Clock Plug-In for Data Items**
Visibility Level: framework-level
Adaptation Time: run-time
Adaptation Method: plug-in component in `DataItemWrite` class
Pre-defined Options: none
Related Hot-Spots: AOCS clock plug-in for `AocsObject` class
Description `DataItemWrite` objects attach a time tag to each value they write. They therefore need a clock to provide them with the time. Class `DataItemWrite` offers method `setAocsClock` to define the plug-in clock that is used for this purpose. Note that the link to the clock component is static and hence the clock only needs to be plugged into one instance of class `DataItemWrite`.

This table is taken from the concept and architecture description document of the intercomponent communication framelet in the AOCS Framework.

7 The User's Perspective

The users alluded to in the title of this chapter are the users of the framework, namely the application developers who are planning to use the framework to help them develop their applications. Previous chapters considered software frameworks from the point of view of their designers. This one takes the point of view of their users.

7.1 A Reuse-Driven Development Process

The top half of figure 7.1 shows a simplified view of a typical development cycle as it is adopted in the development of single applications. The development process is driven by a specific need. It is this need that gives rise to the system requirements from which all design decisions originate and to which all design constructs must be ultimately traceable. This driving need is represented by a "customer" who is the entity procuring the software. The "supplier" is the entity responsible for producing and delivering the software. The customer/supplier split is very rigid and, at least in theory, the flow of control is entirely from left to right. Software reviews – the main form of interaction between supplier and customer while the development is under way – always assign the active role to the entity in charge of the system specification. This approach is of course perfectly *compatible* with software reuse since there is nothing to prevent the supplier from reusing existing components in an attempt to reduce costs or schedule. Reuse however might be coming into play too late because by tailoring the system requirements to the specific needs of the project at hand, the likelihood of finding existing software artifacts that match these requirements is very small. The result of this approach is an application that is highly optimized for the purpose for which it was built but that probably had to be entirely handcrafted.

Additionally, mere compatibility may not be sufficient to make software reuse more pervasive in practice. In order to overcome the psychological and cultural barriers that currently relegate reuse to a marginal role, a more proactive role where reuse is explicitly built into the development process may be required. Consider now the diagram at the bottom of the figure. Here, formulation of the system requirements is preceded by a market survey to identify already existing products that might be useful in the current project. If such products are found, the system requirements are written taking them into consideration (albeit implicitly). Responsibility for selecting and using them remains with the software suppliers but

A. Pasetti: Software Frameworks, LNCS 2231, pp. 99-110, 2002.
© Springer-Verlag Berlin Heidelberg 2002

their reuse decisions are made easier by having to comply with requirements that were written with compatibility with existing products in mind. The result of such an approach is a software system that is probably not perfectly tuned to the needs of the project – since it is unlikely that products developed for genericity and reuse will perfectly match the needs of any given project – but that can be probably developed at a lower cost and in less time.

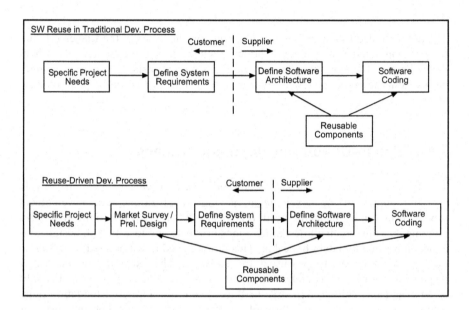

Fig. 7.1. Software reuse in the software development process

It is worth noting that the development process in the bottom diagram is the one normally used for hardware systems where, prior to issuing the system specifications, the customer performs a feasibility analysis that often culminates in the definition of a preliminary system architecture built from pre-existing components. In the case of hardware systems, this way of proceeding is adopted as a matter of course and hardly needs to be justified or explicitly encouraged. Software engineers, by contrast, have a different cultural heritage more oriented towards product customization. Hence, if reuse of software is to become more widespread, its development process must incorporate some measures to actively foster reuse.

Two major requirements follow from the adoption of such a reuse-driven approach. On the one hand, software standards must be updated to explicitly include a pre-specification stage where the customer performs a survey of existing products to identify those applicable to his project. On the other hand, products that are intended for reuse must be such as to make this pre-specification evaluation easy. In the case of frameworks, this means that it should be possible to quickly and effectively match an application with a framework to verify to what extent the latter could be used to generate or help the development of the former. The next sections offers some considerations – falling rather short of a full solution – of how this can be done in practice.

7.2 The Functionality Concept

Prospective users of a framework need a means to carry out a systematic and formalized check of the adequacy of the framework to their application. They cannot use the framework requirements for this purpose because requirements will in general not be specified for a framework (see section 5.1). The formal description of the framework's domain through the corner applications of chapter 5 is also unsuitable as a reference against which to match a target application because the corner applications are used only to initiate the framework specification process which then continues iteratively in parallel to the design process. By the time the framework has been completed, the corner applications may no longer reflect the features it implements. Implementation cases can serve in lieu of requirements to specify the framework and could thus in principle be used to verify compatibility of an application with the framework. The description of a framework that they offer is however indirect and cannot be easily matched to the requirements of an application.

In order to allow evaluation of the suitability of a framework for a given application, a new concept is required that shifts the focus from *a priori* specification of a piece of software that does not yet exist, to *a posteriori* description of a software that has already been created. The emphasis on *a posteriori* description is a consequence of the iterative nature of the framework specification process. Conventional applications are developed to match a (supposedly) fixed set of specifications. Consequently, their specifications offer a full and formal description of the application. In the case of frameworks instead, specifications are fluid and evolve during the design process. In practice, they may never reach the state of maturity that is required for a formal characterization of the framework. It is this lack of precision at specification level that creates a need for a formal *a posteriori* description of the framework. We have coined the term *functionality* to designate this new concept. In the AOCS Framework project, we have tried to give a concrete contour to this concept to make it of practical value to other projects. Below, the functionality concept is described as it emerged form these experiments but their results should be considered as tentative and in need of confirmation and further refinement.

The operational definition of functionality that we adopted in the AOCS project is: an operation or a set of related operations that the framework offers either as fully implemented features or as features potentially implemented by a hot-spot.

Functionalities are intended for persons who are not familiar with the framework design. They should describe the framework from the point of view of a potential user. In this sense, they can be seen as requirements written a posteriori. Their formulation must be made without reference to design items that would normally not be known to potential framework users. In practice concise formulation requires some kind of domain vocabulary. For this purpose, we propose that the domain dictionary constructed during the framework concept definition phase be used. This assumes that potential framework users are familiar with the domain dictionary and with the framework domain model. The assumption is reasonable as potential framework users are by definition developers of applications that lie within the framework domain. Thus, in summary, a functionality can be defined as a description of a capability – either actual or potential – offered by a framework to potential users who are unacquainted with the framework design but are familiar with the framework domain abstractions and model.

Note that the concept of functionality only covers the *functional* aspects of a framework and of the applications that can be instantiated from it. Non-functional aspects such as reliability, performance, inter-operability, etc, are not covered.

A terminological caveat is in order here. The term "functionality" is also used elsewhere in this book in expressions like "functionality manager". The two usages are not related. In the second case, the term "functionality" designates a bundle of related operations that are seen as a single entity. The functionality manager, for instance, is a component that is responsible for controlling a set of related operations. The context will make clear how the term is to be understood.

7.2.1 Functionality Types

Applications can be simply described in terms of what they *do*. Describing a framework is a more complex undertaking because a framework is not a working application and it is defined not only by what it does but also by what it *could* do *if* it were appropriately customized or *if* suitable components were plugged into it. In order to capture this greater range of possibilities, three basic types of functionalities for frameworks have been defined: do-functionalities, can-functionalities and offer-functionalities. They are described in table 11.

Table 11. Functionality types

Functionality Type	Description
Can-Functionality	A can-functionality expresses a capability that is potentially offered to applications derived from the framework but that is not uniquely defined by the framework. Typically, can-functionalities correspond to hot-spots in the framework.
Do-Functionality	A do-functionality expresses a capability that is offered as a fully implemented feature by the framework and that is not intended to – or cannot – be overridden or customized by derived applications.
Offer-Functionality	An offer-functionality expresses a capability that is fully implemented by the framework but that is offered as an option and that applications may choose to use or to ignore. Typically, offer-functionalities correspond to plug-in components offering default implementations of can-functionalities.

The possibility was also considered of having an additional kind of functionality describing what a framework *cannot* do. However, the experience from applying the functionality concept to the AOCS Framework indicates that do-functionalities are sufficient to demarcate the framework.

The formulation of functionalities could in principle be left completely free but this would make their formal treatment impossible. Formal treatment is desirable as it makes it possible, at least in principle, to provide automatic tools that can check coverage completeness or that can help in visualizing functionalities, their mutual relationships and their mapping to application requirements and to design features. The recognition of distinct functionality types imposes some structure on the set of functionalities and opens the way to their formal treatment. Formal treatment is further promoted by introducing *constrained relationships* between different functionalities and, as described in the next subsection, *constrained mappings* from the functionalities to the architectural constructs.

Functionalities can enter into relationships with each other. A functionality can *expand* another functionality, or it can *implement* it, or it can *use* it. These relationships are described in table 12.

Table 12. Relationships between functionalities

Functionality Relationship	Description
Fx expands-to Fy	The expands-to relationship exists between one high-level functionality (Fx) and several lower-level functionalities (Fy) that collectively represent the high-level functionality. This relationship typically arises when a complex functionality is described in terms of simpler functionalities.
Fx is-implemented-by Fy	This relationship implies that Fy describes an implementation of Fx. It typically arises when a framework offers a component (embodying an offer-functionality) as a default plug-in for a hot-spot (embodying a can-functionality).
Fx uses Fy	This relationship arises when functionality Fx needs functionality Fy. This relationship exists when the implementation of a functionality relies on other functionalities.

Functionality relationships must satisfy some constraints that follow from their semantics and from the semantics of the functionality types. A constraint on a relationship "Fx *relationship* Fy" limits the types of functionalities Fx and Fy. The following constraints apply to relationship "Fx *expands-to* Fy":

- if Fx is a can-functionality, then Fy can only be a can-functionality or a do-functionality
- if Fx is a do-functionality, then Fy can only be another do-functionality
- if Fx is an offer-functionality, then Fy can only be another offer-functionality

A single constraint applies to relationship "Fx *is-implemented-by* Fy":

- Fx must be a can-functionality and Fy must be an offer-functionality representing a default implementation of the can-functionality.

No constraints instead apply to the *uses* relationship.

7.2.2 Mapping Functionalities to Architectural Constructs

Mappings from functionalities to architectural constructs describe how the functionalities are implemented by the framework. They allow functionalities to be traced to the framework architectural constructs. Software development processes for individual applications normally demand that requirements be traceable to architectural features to allow formal verification of their correct implementation. When an application is instantiated (either in part or wholly) from a framework, some of the framework constructs are incorporated into it. Some framework components, for example, become application components. Some of the application requirements will therefore be directly implemented through the framework. Hence, traceability from requirements to architectural features is only possible if there is traceability from the requirements to the framework functionalities and from the latter to the architectural features. Traceability from requirements to functionalities is discussed in the next section. This section considers instead traceability from functionalities to architecture.

Traceability to architecture in an application context normally means traceability to concrete operations exposed by application components. Frameworks, however, embody more abstract functionalities which cannot necessarily be mapped to concrete constructs. Consider for instance a can-functionality that expresses a potentiality, something that the framework code could do if it were properly customized, for instance, by subclassing an abstract class and providing an implementation for an abstract method. Or consider a functionality that describes a design pattern exported by a framework. In both case, it is not possible to identify a concrete component or method that implements the functionality. Such functionalities must be mapped to other kinds of architectural constructs.

Some latitude is possible in selecting the types of architectural constructs to which functionalities can be mapped. In the AOCS Framework project, we only restricted functionalities to be mapped to the following constructs:

- Hot-Spots
- Exportable Design Patterns
- Components

There is no mapping to abstract interfaces because abstract interfaces always give rise to a corresponding hot-spot. Design patterns can be targets of functionalities because design patterns are exported by the framework to applications – they are part of the framework's "offering" – and must therefore embody some of its functionalities. However, design patterns that are instantiated internally to the framework are matched by hot-spots since an instantiated design pattern exists in order to model hot-spot adaptability. Exportable design patterns instead act as carriers for functionalities that cannot be further reduced to other constructs.

This mapping philosophy favours hot-spots as targets for functionalities over abstract interfaces and (instantiated) design patterns because they are usually closer to concrete constructs and thus more suitable for verification purposes. Hot-spots are also more understandable to application developers who are the target audience for the functionalities.

If a functionality cannot be mapped to one of the constructs listed above, this is an indication that it needs to be broken down into lower-level functionalities.

Two types of mappings have been identified between a functionality and an architectural construct: *is-implemented-by* and *matches*. The meaning of the mapping depends on the type of the functionality and on the type of the architectural construct. The following pairs with the following meanings are possible:

− *CFx matches HSy*: This mapping arises between a can-functionality CFx and a hot-spot HSy. The can-functionality represents a capability that is potentially offered by the framework. HSy is the hot-spot where the framework behaviour is adapted to make the potential functionality represented by CFx actual.

− *OFx matches HSy*: This mapping arises between an offer-functionality OFx and a hot-spot HSy . An offer-functionality represents a capability that is offered as an option by the framework and that typically represents a default implementation of a can-functionality. HSy represents the hot-spot where the behaviour adaptation implied by the offer-functionality is achieved.

− *DFx is-implemented-by Core-Component/Exportable Design-Pattern*: This mapping arises between a do-functionality DFx and a core component or an exportable design pattern exported by the framework. A do-functionality represents an actual capability of the framework. The core component or the design pattern represent the constructs that implement the capability.

− *OFx is-implemented-by Default-Component*: This mapping arises between an offer-functionality OFx and a default component exported by the framework. An offer-functionality represents a capability offered by the framework as an option. The default component represents the construct that implements the capability.

The following constraints apply to the mapping between a functionality and an architectural feature:

− A can-functionality or a hot-spot can only enter into one *matches* mapping
− An offer-functionality can only enter into one *matches* mapping.
− if mappings "CFx is-implemented-by Ofy" and "CFx matches HSz" hold, then the following mapping must also hold: "OFy matches HSz".

7.2.3 Completeness of Description

Given a set of functionalities that purport to describe a framework the obvious question that arises is whether the set is *complete*. Intuitively, a complete set of functionalities is one where all functionalities can, either directly or indirectly, be mapped to the architectural constructs of the framework and where all architectural constructs are covered by at least one functionality. In order to formalize this intuitive completeness criterion, the following definitions are required:

- A functionality F is *directly mapped* to an architectural construct C if either of the following mappings hold: (F matches C) or (F is-implemented-by C)
- A functionality F is *indirectly mapped* to an architectural construct C if the following chains of relationships and mappings hold:

$$(\text{F relationship } F_1)$$

$$(F_1 \text{ relationship } F_2)$$

$$(F_2 \text{ relationship } F_3)$$

$$. \quad . \quad .$$

$$(F_{N-1} \text{ relationship } F_N)$$

$$(F_N \text{ mapping } C)$$

where relationship `relationship` is either the *is-implemented-by* or the *expands-to* relationship or the *use* relationship, and mapping `mapping` is either the *matches* or the *is-implemented-by* mapping.
- An architectural construct is *covered* by a set of functionalities if there is at least one functionality in the set that can be either directly or indirectly mapped to the construct.

Given the above definitions, then the intuitive criterion for completeness can be more formally formulated as follows: a set of functionalities for a framework is complete if all the functionalities in the set can be either directly or indirectly mapped to a framework architectural construct and if all the framework architectural constructs are covered by the set of functionalities.

7.2.4 Mapping Requirements to Functionalities

With the reuse philosophy outlined at the beginning of this chapter, in the initial phase of a project, the application designers must decide whether or not to use a given framework to assist them in the development of their application. The first step in such a decision must be to establish whether the application lies within the framework domain. In the AOCS Framework project, we considered how this can be done in the case of an application that is described by a set of user requirements and of a framework that is described by a set of functionalities. The basic idea is to match the requirements to the functionalities on the assumption that if the application is to be generated from the framework, then all, or at least a large number, of its requirements should be covered by one or more functionalities.

In order to verify whether this is the case, the concept of *requirement mapping* is introduced. More specifically, three types of mappings from application requirements to framework functionalities are defined:

- is-compatible-with
- is-implemented-by
- is-incompatible-with

The *is-compatible-with* mapping can only arise between an application requirement R and a framework can-functionality CFx. The mapping "*R is-compatible-with CFx*" expresses the fact that CFx describes a potential capability of the framework that, if it were implemented, would represent an implementation of the requirement. It must be stressed that since the can-functionality represents a *potential* capability, the existence of the mapping simply says that the requirement is *compatible* with the framework. It does not imply that the requirement is actually implemented by the framework.

The *is-implemented-by* mapping arises between an application requirement R and a framework functionality Fx where Fx is either a do-functionality or an offer-functionality. The mapping "*R is-implemented-by Fx*" expresses the fact that Fx describes a functionality that represents a direct implementation of the requirement.

Finally, the *is-incompatible-with* mapping arises between an application requirement R and a framework do-functionality DFx. The mapping "*R is-incompatible-with DFx*" expresses the fact that Fx describes a functionality that is incompatible with the requirement.

Given an application and a framework, the coverage analysis is done by analyzing all the application requirements and mapping them to the framework functionalities. Three cases are then possible. If the application is wholly inside the framework domain, then it must be possible to establish *is-implement-by* mappings between all its requirements and the framework functionalities. This implies that the application can be entirely generated by the framework using its pre-defined constructs. If the application is only partially inside the framework domain, then none of its requirements must have an *is-incompatible-with* mapping to any of the framework functionalities. In this latter case, the degree of coverage of the application can be measured by two metrics:

- the proportion of the application requirements that have a mapping either of the *is-implemented-by* or of the *is-compatible-with* kind with the framework functionalities.
- the ratio between the number of requirements having an *is-implemented-by* mapping to the framework functionalities and the number of requirements having an *is-compatible-with* mapping to the framework functionalities.

Finally, the third case arises when some of the requirements have an *is-incompatible-with* mapping to the framework functionalities. This places the application outside the framework domain.

The various mappings and functionality relationships introduced in this chapter create a web of interconnections among application requirements, framework functionalities and framework constructs. One could imagine to represent this web graphically as in the example in figure 7.2. The dashed arrows represent the mapping from the functionalities to the framework constructs, the dotted arrows represents the relationships among the functionalities themselves and the solid arrows show the mappings from the application requirements to the functionalities. This type of diagram gives graphical form to a traceability matrix that shows how a framework can help implement the requirements of a given application.

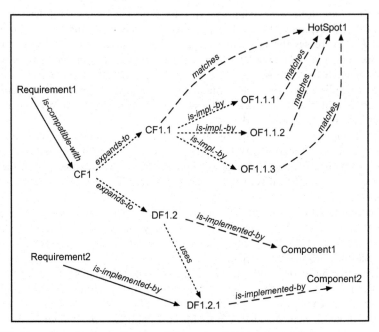

Fig. 7.2. Map of requirements, functionalities and architectural constructs

7.2.5 Functionalities in the AOCS Framework

The functionality concept was elaborated in the AOCS Framework project. Functionality lists were set up for some of the framelets but, unfortunately, there never was an opportunity to test their adequacy by matching them to the requirements of a "real" AOCS system. The validity of the classification of functionality types and of their relationships to each other and to architectural constructs and application therefore remains somewhat in doubt. However, the functionality concept itself – the formal description of a framework after it has been developed – is regarded as sound and as essential to the effective adoption of the framework concept in concrete application development projects.

By way of example, chapter 21 presents a functionality list for one of the framelets of the AOCS Framework. The list is incomplete but it nonetheless serves to illustrate the concepts presented in this chapter and to provide a template for the definition of functionalities for other frameworks.

7.2.6 Alternative Approaches

As already noted in the introduction to this chapter, the definition of the functionalities, in some sense, replaces domain analysis and it is therefore not surprising that a very similar concept comes precisely from this field. The FORM method

[55, 56] is introduced to discover and capture commonalities among applications in the same domain. Domain description is based on the concept of *feature,* namely "distinctively identifiable functional abstractions that must be implemented, tested, delivered and maintained" [55, p. 144]. Interestingly, the introduction of features as a descriptive vehicle for domains is motivated by the fact that customers and engineers typically speak of their products in terms of the features. Thus, features, like functionalities, are created with an eye to the needs of the users of the reusable software artifacts that they describe.

This is not the only similarity between features and functionalities. Features have types, they can be "mandatory", "optional" or "alternative", which resemble the functionality types identified above. Features can enter into three types of relationships with each other of which two – "composed of" and "implemented by" – have an obvious equivalent in the functionality relationships of this section. Features diagrams are also proposed that are conceptually identical to those of figure 7.2.

Features, like functionalities, should be mappable to architectural elements. In a difference with functionalities, however, the target architectural elements of features seem to be only components and their methods whereas functionalities can also be mapped to design patterns and hot-spots. Functionalities therefore seem to take better account of the specific nature of frameworks (at least of frameworks as they have been defined in this work).

As in the case of functionalities, mapping to architectural elements is the precursor to matching to application requirements and [55] recommends that the development of a new application should begin with a process of systematic matching of the application requirements to the domain features. This activity is similar to the requirements coverage analysis discussed above.

Another reference that proposes a feature-based domain analysis along similar lines is [20]. The scoping problem is explicitly defined as the problem of selecting the features to be included in a framework where a feature is a logical unit of behaviour specified by a set of functional or quality requirements. As in reference [55], formal relationships can exist between features and "feature graphs" are put forward as a convenient domain analysis tool.

In an important respect, features – as defined and used by both the above references – are more comprehensive than functionalities because they can also model the non-functional aspects of an application (e.g. reliability, real-time performance, inter-operability issues, etc). During the application development process, non-functional features are used to select among alternative components and to define the component connections. This is a useful capability and one that functionalities definitely lack. The fundamental difference between features and functionalities, however, lies in the role they play in the overall framework development process. Features are defined *prior* to the framework development and are an input to it. Functionality definition instead concludes the framework development process. In practice, the formulation of the functionalities will also be easier than that of features because it is easier to describe an artefact that exists (albeit at a high level of abstraction) than it is to characterize *a priori* a domain. Thus, as far as application development is concerned, functionalities are a more efficient tool than features but on the other hand features were proposed with a far more ambitious goal since they are to serve during framework development as well as during application development.

8 General Structure of the AOCS Framework

This chapter opens the second part of the book which is devoted to the AOCS case study. Its objective is to show how an object-oriented software framework can be built for a complex embedded control system. The presentation of the case study is the chief purpose of this book and serves to justify the methodological concepts described in earlier chapters since the development of the AOCS Framework acted as both a spur for their introduction and as a test bed for their validation. The material that follows will give several examples of their usage and usefulness.

As already mentioned, a project web site is maintained from which all the code and documentation for the AOCS Framework can be freely downloaded (see footnote on page VI). For those who intend to avail themselves of this option, the warning must be repeated that the AOCS Framework was developed as a proof-of-concept product and its quality and completeness are those of a research prototype.

This book describes the framework mainly at the concept level (in the sense of chapter 6). More specifically, it offers a description that has a level of detail that is intermediate between that which is prescribed for the framework concept definition phase and that which is prescribed for the framelet concept definition phase. The next chapter presents the design principles adopted in the AOCS Framework project. The following chapters cover the individual framelets in which the AOCS Framework was divided. A final chapter addresses the framework instantiation problem. The appendix gives the domain abstraction dictionary and the list of framework design patterns and hot-spots. It can be used as a reference while reading the main body of the book.

8.1 The RTOS Example

Although the design problems that had to be addressed when developing the framework were both numerous and varied, high-level uniformity was preserved by imposing upon the framework a general structure whose description is the chief goal of the present chapter. This will be done by developing an analogy between the AOCS Framework and Real Time Operating Systems (RTOS). This analogy is significant both because it provided the inspiration for the AOCS Framework and because it points to an interesting parallel between frameworks and operating systems leading to a view of frameworks as domain-specific extensions to the operating systems.

A. Pasetti: Software Frameworks, LNCS 2231, pp. 111-129, 2002.
© Springer-Verlag Berlin Heidelberg 2001

One of the challenges of the AOCS Framework project lay in the real-time and embedded character of this type of applications which have seldom been the targets of frameworks. In the initial phase of the project, however, it was realized that there is at least one highly successful example of frameworks for real-time systems. As will be shown in a moment, RTOSs fully fit the definition of framework given in section 3.2 and, because of their record in providing reusable software solutions, it was felt that they could serve as conceptual models for the AOCS Framework. Accordingly, the AOCS Framework project started with an attempt to answer the question: "How does an RTOS achieve reusability?". Three complementary answers were identified. They are illustrated below for the case of task scheduling which is a typical functionality[12] offered by RTOSs.

Firstly, an RTOS enforces an *architectural model*. For instance, it may assume an application to be made up of tasks and it may prescribe certain mechanisms for inter-task communication. Applications that wish to use the RTOS must conform to this model. Using the terminology of section 3.2, the RTOS proposes and imposes a design pattern.

Secondly, the RTOS relies on the *separation of the management of a functionality from its implementation* and achieves this separation through an abstract interface. In the case of task scheduling, for instance, and using UML notation, the RTOS sees a task as a component implementing the following abstract class:

Schedulable
init() : void execute() : void

where a call to init() causes the task to initialize itself and a call to run() causes it to start executing. Separation through an abstract interface fosters reusability because it decouples the implementation of the task, which is necessarily application-specific, from the scheduling policy which is used to activate it and which is application-independent.

Thirdly, the RTOS provides an application-independent and hence reusable component – a core component in the terminology of section 3.2 – that encapsulates the management of the RTOS functionalities. The RTOS executable, for instance, will normally include a scheduler component that can be made application-independent because it sees the tasks it manages only through the abstract schedulable interface. The conceptual class structure of the scheduler is shown in figure 8.1.

Thus, the RTOS offers all three frameworks constructs identified in section 3.2 and qualifies as a framework because the design pattern and the abstract interfaces are introduced to model adaptability and the component packages the invariant part of RTOS-based applications. It may be noted in passing that the primary entities are the design pattern and the schedulable abstract interface. The scheduler

---

[12] The term "functionality" here designates a bundle of related operations that are treated as a single conceptual entity. This term was used with a different meaning in chapter 7.

component is secondary as its function is essentially defined once the design pattern and abstract interface have been defined.

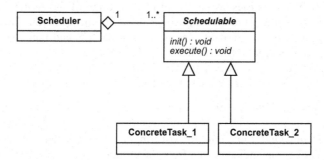

Fig. 8.1. Conceptual class structure of a scheduler

Figure 8.2 gives an alternative view of an RTOS-based system. The RTOS module is shown as a component exposing slots into which application-dependent components can be plugged. The application-dependent components represent the tasks to be managed by the RTOS. The slots in the RTOS are therefore the hotspots where the RTOS customization takes place. The "shape" of the slots that determines whether a certain component can be plugged into it is defined by the abstract interface that separates task management from task implementation.

Finally, it hardly needs to be mentioned that RTOSs are not built using object-oriented languages and the analogy described above is conceptual rather than factual. It should however be noted that RTOSs achieve reusability and extensibility through the call-back style of programming which is typical of frameworks [41].

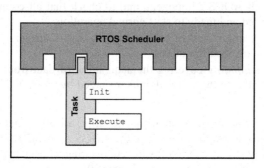

Fig. 8.2. The RTOS scheduler as component customizable by plug-in tasks

8.2 The Lesson for the AOCS

One conclusion of the analysis in the previous section is that in an AOCS there is at least one functionality – task scheduling – where a framework approach can improve reusability. Task scheduling is of course only one of the many functionalities implemented by an AOCS and it is therefore natural to ask whether the same principles that make scheduling management reusable could be applied to the other AOCS functionalities. A large part of the design of the AOCS Framework can be seen as an attempt to give a positive answer to this question. This was done by first isolating the functionalities present in an AOCS and then systematically applying to each the reusability model of the RTOS.

The functionalities that were identified in an AOCS are listed below. Chapter 2 provides the necessary domain background and justification for their introduction:

- The *Telemetry Functionality* covering the formatting and despatching of housekeeping information to the ground station.
- The *Telecommand Functionality* covering the management, processing and execution of commands received from the ground station.
- The *Failure Detection Functionality* covering the management of checks to autonomously detect failures in the AOCS software.
- The *Failure Recovery Functionality* covering the management of the corrective measures to counteract the failures reported by the failure detection functionality.
- The *Controller Functionality* covering the management and implementation of the control algorithms for the control loops operated by the AOCS.
- The *Manoeuvre Functionality* covering the management of manoeuvres, namely coordinated sequences of actions performed by the on-board software with a specific goal.
- The *System Functionality* covering the management of system-wide operations such as the reset function that bring the AOCS components to some initial default state or the configuration control operations to clear all configuration information in all AOCS component, and to check that all components are correctly configured and ready to start normal operation.
- The *Reconfiguration Functionality* covering the management of redundant units and of their reconfigurations.
- The *Unit Functionality* covering the acquisition of data from and the forwarding of commands to the external units managed by the AOCS (the sensors and the actuators).

With the exception of the system functionality, all other functionalities in the list can be directly traced to features of the AOCS domain. The need for the system functionality arose from software engineering considerations to facilitate application reset and initialization (see chapter 10).

From the point of view of the framework designer, the functionalities can be treated as independent of each other. This is obvious in all cases with the possible exception of the failure detection and failure recovery functionalities that in some systems are merged together. In the AOCS Framework, however, the failure detection manager constructs failure reports encapsulating the description of any failures it has encountered and deposits them in shared repositories. In a separate activity, the failure recovery manager inspects the failure repository and processes the failures according to their description.

The mutual independence of the functionalities was crucial in simplifying the framework design as it allowed architectural solutions to be developed in isolation for each functionality. Architectural independence, however, did not degenerate into arbitrariness of solutions. As argued in section 8.5, design unity was maintained by imposing the same meta-pattern on all functionality management problems.

The influence of the RTOS example on the AOCS Framework design will now be illustrated by retracing the steps that led to the definition of an architectural solution for the telemetry and controller management problems. This description anticipates that of chapters 21 and 22 documenting the telemetry and controller framelets.

8.3 Telemetry Management in the AOCS Framework

In current AOCS systems, telemetry processing is controlled by a so-called telemetry handler that directly collects telemetry data, formats and stores them in a dedicated buffer, and then has them transferred to the ground. To accomplish its task, the telemetry handler needs an intimate knowledge of the type and format of the telemetry data: it has to know which data, in which format, and from which objects they have to be collected. It is this coupling between telemetry handler and telemetry data that makes the former application-specific and hinders its re-use.

In keeping with the RTOS reuse model, the approach taken in the AOCS project is based on a design pattern – the *telemetry design pattern* – that calls for a separation of telemetry management from the implementation of telemetry data collection. This is achieved by endowing selected components in the AOCS software with the capability of writing their own state to the telemetry stream. Telemetry processing is then seen as the forwarding to the ground of an image of the internal state of some of the AOCS components. The resulting architecture is shown in figure 8.3.

The ability of a component to write its own state to the telemetry stream is encapsulated in the abstract interface Telemeterable which must be implemented by all components that contribute state information to the telemetry data. Its basic method is writeToTelemetry. Calling it causes a component to write its internal state (or a subset of it) to the telemetry stream. The telemetry manager is responsible for keeping track of the components whose state should be sent to the telemetry stream, for calling their writeToTelemetry methods, and for starting the forwarding of telemetry data to the ground. The telemetry manager also needs to be aware of the data stream to which the telemetry data should be written. Since the hardware characteristics of the channel over which telemetry data are transmitted to the ground differ across AOCS systems, the telemetry stream is identified by an abstract interface (interface TelemetryStream in the figure). This interface defines the generic operations that can be performed on a data sink representing an ideal telemetry channel and decouples telemetry management from hardware implementation details.

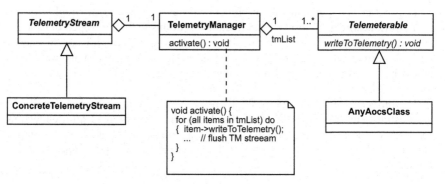

Fig. 8.3. Telemetry management in the AOCS Framework

The second step in the development of an architectural solution for AOCS telemetry management was the instantiation of the telemetry pattern for the AOCS domain. This chiefly involved providing exact definitions for the `Telemeterable` and `TelemetryStream` interfaces. The former, for instance, was eventually defined as follows:

Telemeterable
writeToTelemetry(TelemetryStream) : void *setTelemetryFormat(int) : void* *getTelemetryFormat() : int* *getTelemetryImageLength() : int*

In the selected implementation, the telemetry stream is passed as an argument to method `writeToTelemetry` and methods are provided to check the size and to set the format of the telemetry image generated by each component.

In the third and final step of the architecture design process, the telemetry manager component was characterized. The semantics of the telemetry pattern and of its associated abstract interfaces make its definition straightforward. The core of the telemetry manager component can be represented in pseudo-code as follows:

```
Telemeterable*  tmList[N];
TmStream*       tmStream;
. . .
void activate()    {
   for (all telemeterable objects t in tmList) do
   {  t->setTelemetryFormat(tmFormat);
      tmDataSize=t->getTelemetryImageLength();
      if (object image fits in TM buffer)
         t->writeToTelemetry(tmStream);
      else
         . . .      // error!
   }
}
```

Thus, the telemetry manager maintains a list of telemeterable components and, when it is activated, it calls their `writeToTelemetry` method. It uses the other operations exposed by `Telemeterable` to control the format of the telemetry images and to verify that the size of their image is compatible with the capacity of the telemetry channel. Additionally, and not shown in the pseudo code segment, the telemetry manager will offer operations to dynamically load and unload items from its internal list of telemeterable components and to set the telemetry stream at initialization. Clearly, all these operations are application-independent and hence the telemetry manager is fully reusable or, in other words, it becomes a core component in the framework. The AOCS Framework also provides a default component implementing the `TelemetryStream` interface for the case of a DMA-based telemetry channel that happens to be common in AOCS systems.

The architectural solution to the telemetry management problem therefore exhibits the typical features of a framework being based on a design pattern that is specialized by abstract interfaces and a core component (the telemetry managers) and complemented by a default component (the default implementation of interface `TelemetryStream`).

Figure 8.4 shows a plug-in view of telemetry management. It again draws attention to the similarity between the telemetry management and the RTOS scheduling architectures and on how in both cases reusability originates in the introduction of an abstract interface to separate the management of a functionality from its implementation.

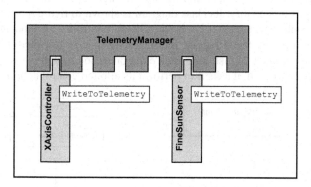

Fig. 8.4. The telemetry manager as a component customizable by plug-in components

8.4 Controller Management in the AOCS Framework

A *controller* is a component that encapsulates a closed-loop controller. Closed-loop controllers in AOCS applications are typically used to stabilize the spacecraft attitude and its orbital position. They tend to implement the same flow of activities: measurements from sensors are acquired, they are passed through a compensation filter, and commands are computed and applied to actuators to counteract any deviation of the variable under control (e.g. the satellite attitude) from its desired value.

The first step in the design of a generic solution for the controller functionality in the AOCS Framework is the definition of a *controller abstraction* that plays the same role as the *schedulable* abstraction in the RTOS case and the *telemeterable* abstraction in the telemetry management example. This abstraction is represented by an abstract interface that defines the generic operations that can be performed on a closed-loop controller independently of the particular algorithms that it implements:

Controllable
doControl() : void *isStable() : bool*

Interface `Controllable` must be implemented by all components representing closed-loop controllers. It recognizes two operations. `doControl` directs the component to acquire the sensor measurements, derive discrepancies with the current set-point, and compute and apply the commands for the actuators. Since closed-loop controllers can become unstable, an additional operation, `isStable`, is provided to ask a controller to check its own stability.

The controller abstraction separates the implementation of the control algorithms from their management and allows a design pattern to be constructed for the controller functionality. Its class diagram is shown in figure 8.5. The pattern is based on a controller manager that holds a list of controller components which are seen through the `Controllable` interface. When it is activated, the controller manager goes through the list, asks each component to check its own stability and, if stability is confirmed, it asks it to perform its allotted control action.

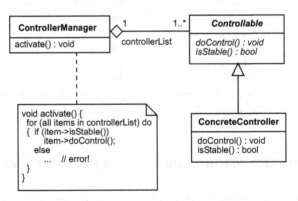

Fig. 8.5. Class diagram for controller management in the AOCS Framework

In terms of the definitions of section 3.2, the controller manager is a *core component* because its implementation is application-independent and it can therefore be reused "as is" in any AOCS application instantiated from the AOCS Framework. The concrete controller classes of the controller pattern are instead application-specific (each closed-loop on each satellite has its own control algorithm). However, there are some types of control algorithms that recur often. It therefore makes sense for the framework to provide some default implementations of interface `Controllable` that can be configured to implement them. In the terminology of section 3.2, these are *default components*. Application developers then have a choice to provide their own entirely new implementation, to use a default implementation, or to modify a default implementation by subclassing it and overriding selected methods.

This solution – like the solution to the telemetry management problem – has the same structure as the RTOS solution and it conforms to the definition of frameworks given in section 3.2. A plug-in view of the controller manager is also possible and is shown in figure 8.6.

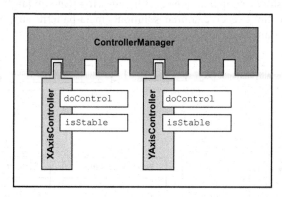

Fig. 8.6. The controller manager as a component that can be customized by plugging into it components implementing interface `Controlable`

The solution sketched above is at the design level only. A realistic implementation will have a richer set of operations for the `Controllable` interface and for the controller manager component. These operations are defined when the controller pattern is instantiated during the architectural definition phase. It is only in this phase that the external interfaces of the components are fully defined.

Finally, note that, as in the case of the telemetry management problem, the design problem centres on the definition of the controllable abstraction and of the controller design pattern. Once these have been defined, the operations to be offered by the concrete components – the controller manager and any default controllers – and their implementations follow naturally. This again confirms the view, repeatedly put forward in the first part of this book, that design patterns and interfaces are the key concepts in framework design.

8.5 The Manager Meta-pattern

The procedure described above for the telemetry and controller functionalities was followed for all other functionalities identified in the AOCS. In all cases, architectural solutions were produced that were conceptually similar to that for the telemetry functionality and that were inspired by the RTOS model of reusability. This commonality of design solution is so significant that it deserves to be captured in a new concept: the *manager meta-pattern*.

Design patterns define solutions for design problems at application level: they prescribe recipes that allow a class of related design problems confronting application developers to be solved in a uniform manner. The term meta-pattern is instead proposed to describe commonalities in different design patterns used to address different design problems in the same framework. Just as design patterns promote consistency at application level by ensuring that similar problems arising in different locations receive similar solutions, meta-patterns support design consistency at the framework level[13].

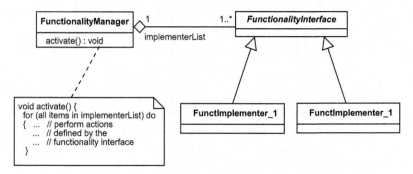

Fig. 8.7. The manager meta-pattern of the AOCS Framework

The manager meta-pattern (see the class diagram of figure 8.7) addresses the problem of managing a functionality that requires the same actions to be repeatedly performed on a class of components providing different implementations of the functionality itself. The solution it proposes can be summarized as follows:

– Define an abstract interface capturing the behavioural signature of the functionality. This interface separates the management of the functionality from its implementation.
– Build an application-independent and reusable functionality manager component (a core component). The functionality manager maintains a list of components seen as instances of the functionality interface and it exposes an activati-

---

[13] The usage of the term "metapattern" is therefore rather different from that of references [76, 77, 51].

on method that systematically performs all the actions required by the functionality management on all components in the list
– Where appropriate, provide default implementations of the functionality interface (default components)

The manager meta-pattern was applied to all the AOCS functionalities identified in section 8.2. This resulted in a framework architecture that is both elegant and homogeneous. Extensions and upgrades are facilitated by the mutual independence of the functionality managers that makes addition of facilities to handle new functionalities easy.

8.6 Overall Structure

Systematic application of the manager meta-pattern leads to a framework that will generate AOCS applications with an architecture as shown in figure 8.8. A collection of functionality managers represent the reusable architectural skeleton of the application. This is customized for a specific application by composition with components encapsulating the application-dependent functionality implementations. To each functionality manager an abstract interface is associated. The RTOS could be one of the functionality managers and the framework can therefore be seen as offering a *domain-specific extension to the operating system.*

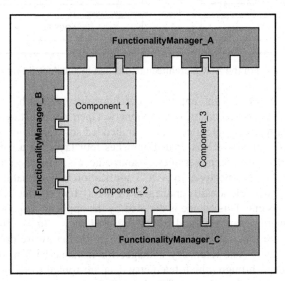

Fig. 8.8. Overall structure of the AOCS Framework

The same component can be used to customize several functionality managers. A closed-loop controller, for instance, would obviously be plugged into the controller manager but might also be plugged into the telemetry manager if it is desired to include its state in the telemetry stream. This situation is handled by having a component implement all the interfaces associated to the functionality managers it customizes. `Component_1` in figure 8.8, for instance, would have to implement (at least) the abstract interfaces corresponding to the functionality managers A and B.

In a language like C++ that does not have the interface concept, implementation of the architecture of the figure requires the use of multiple inheritance. Use of multiple inheritance is often resisted because it can lead to ambiguities but the AOCS Framework only requires multiple inheritance of abstract interfaces which is known to be safe. This kind of multiple inheritance is used extensively throughout the AOCS Framework with typical AOCS components normally implementing 5 or 6 different abstract interfaces each representing a different functionality.

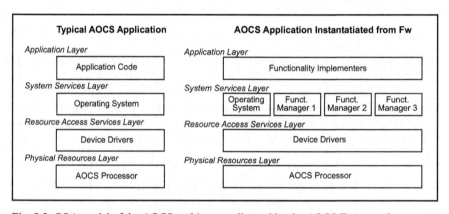

Fig. 8.9. GOA model of the AOCS architecture dictated by the AOCS Framework

Figure 8.9 shows how the application architecture implied by the AOCS Framework can be recast according to the Generic Open Architecture (GOA) model [96]. GOA is a reference architecture for embedded software to provide a foundation for effective open systems. It distinguishes four layers in a software system: application software, system services, resource access services, and physical resources. In a GOA perspective, an AOCS application is most naturally decomposed by placing the functionality managers in the system services layer alongside the operating system. This choice is dictated by the application-invariant character of the functionality managers and by the fact that they – unlike the components implementing the functionalities – need access to the hardware drivers in the resource access layer. The telemetry manager for instance has an interface to the hardware channel through which telemetry data are routed to the ground. The figure illustrates again the affinity of functionality managers and operating systems and shows that the framework factors out of an application OS-like functionalities encapsulating them into application-independent modules.

8.7 Architectural Infrastructure

The functionality managers address localized design issues in the AOCS domain but do not exhaust the framework since they cannot handle non-localized infrastructure problems. Four types of infrastructure problems were identified in the AOCS domain.

Firstly, the architecture promoted by the AOCS Framework is component-based. Some mechanism is required to allow these components to exchange information. The second infrastructure problem arises from the need of components to monitor each other both in order to synchronize their behaviour and for purposes of failure detection. An AOCS application normally has to implement data processing algorithms (e.g. filters, compensators, state estimators, etc). It is usually convenient for the AOCS designer to describe these algorithms as made up of several linked and nested sub-algorithms. The third infrastructure problem accordingly addresses the need to provide an encapsulation of data processing algorithms that allows them to be handled as composable blocks. Finally, the AOCS software is usually operational mode dependent (in the sense of section 2.1.2) and therefore a mechanism is required to allow the AOCS components to change their behaviour in accordance with changes in their environment.

As will be seen in the remainder of this report, the AOCS Framework provides solutions to all of the above infrastructure problems. These solutions are qualitatively different from those proposed for the AOCS functionalities (they are not based on the manager meta-pattern) but they resemble them in being based on the definition of domain-specific design patterns and of the abstract interfaces required to specialize them. In many cases, core and default components were also developed but in the framework design process they always came second to the design patterns and abstract interfaces.

8.8 Hierarchies of Design Patterns

The introduction of the manager meta-pattern points to the existence of hierarchies of design patterns. The manager meta-pattern could itself be seen as an instance of a more general design pattern like the strategy pattern of [44]. Since the strategy pattern also provides a foundation for some of the infrastructure design patterns in the AOCS Framework, one could build a hierarchical tree of the patterns used in the framework of which a few branches are shown in figure 8.10. The notation in the figure is informal. The arrows indicate a relationship of refinement or specialization of design patterns across levels of abstraction and generality.

Two clarifications are in order with reference to the discussion in section 4.1 about the way in which framelets subdivide the space of design patterns. Firstly, the *framelets partition only the domain-specific design patterns that are offered by the framework*, namely those that appear at the bottom of the tree in figure 8.10. There is therefore a fundamental difference between the framework design patterns and higher order patterns like the manager meta-pattern or the strategy pattern.

The framework design patterns are intended to be *directly used* either within the framework (instantiated design patterns) or within the applications instantiated from it (exportable design patterns). The meta-patterns are instead provided simply as an aid to understand the structure of the framework. They should not be directly instantiated in either the framework or in applications derived from it.

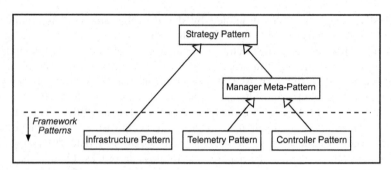

Fig. 8.10. Hierarchies of design patterns in the AOCS Framework

The second point to note is that the labeling of a particular design pattern as a meta-pattern is relative to the design choices made in the AOCS Framework. It is perfectly conceivable that a different design team might have come up with a different framework for the AOCS domain that has the manager meta-pattern as a first-order pattern that is directly offered to AOCS application developers. There is, in other words, nothing intrinsic to the AOCS domain that dictates that the manager pattern must be a meta-pattern whereas the controller and telemetry patterns must be first-order patterns. It just so happens that in the AOCS Framework Project the framework was organized in such a manner that the telemetry and controller patterns are simple (or first-order) patterns whereas the manager pattern is a meta-pattern (or second-order pattern).

This point can be made clearer with an analogy from object-based software design. Consider a software engineer who is designing an application that deals with geometric figures. One possible approach is to introduce a class to represent triangles, one to represent squares, one to represent pentagons, etc. In order to simplify the design process, the engineer could decide to partition the set of classes into subsystems and to allocate the triangle, square and pentagon classes to different subsystems. He could then introduce the concept of "polygon" to capture the commonalities of the triangle, square and pentagon classes. With these design choices, the polygon class becomes a super-class that spans subsystem boundaries. Clearly, other design choices would have been possible. One could for instance have introduced the polygon class as a primary class and have had a dedicated subsystem for it. However, once one has decided to introduce the triangle, square and pentagon classes and to allocate each to a dedicated subsystem, then the polygon class becomes a superclass and it is not possible to allocate it to a specific subsystem. Similarly, in the case of the AOCS Framework, given that the telemetry, controller, etc, patterns have been introduced as first-order patterns, then the manager pattern must be a meta-pattern.

The conceptual parallelism between, on the one hand, architecture and subsystems and, on the other hand, design and framelets has often been remarked upon in this book. This parallelism is understood in a very literal sense: design patterns are regarded as the building blocks of framework design in the same sense in which classes are the building blocks of application architectures and, just as there are hierarchies of classes, there can be hierarchies of design patterns. In both cases, the hierarchies are not absolute but are simply the result of the particular choices of the software engineers.

8.9 The Framework Design Process

One of the characteristic features of the design process advocated in chapter 6 is the introduction of a concept definition stage separate from the architectural definition stage. Its objective is to provide solutions at the design level. The design solution outlined in section 8.4 for the controller functionality is an example of a design-level solution and of the kind of design activities that would be performed during the framelet and framework concept definition phases.

The identification of the controller functionality as a distinct functionality within the AOCS domain requiring a dedicated design solution would normally be done in the framework concept definition phase. In this phase, the developer would also identify the control algorithms used by a particular application as a major framework hot-spot. The need for one or more design patterns to cover it would also be recognized at this stage.

In the framelet concept definition phase, the design solution for the controller framelet is worked out to, basically, the level of detail presented in section 8.4. The solution described in this section presents all three features identified in section 6.1 as characteristic of the design level. Firstly, the abstract interfaces and classes are defined in an incomplete manner. Consider for instance, the Controllable interface. At this stage, it features only two methods doControl and isStable but a realistic implementation will have to include other methods to, for instance, switch between open and closed loop operation or to reset the digital filters that implement the control algorithms. These methods are not important at design level where the emphasis is on identifying a solution to the abstract problem of managing controller operation but they definitely need to be defined at architectural level where – by definition – all interfaces must be fully defined.

Secondly, at design level, no effort is made to integrate solutions to different design problems. Section 8.4 considers the controller problem in isolation from other problems. The concrete controllers, for instance, will eventually have to implement other interfaces besides Controllable. Typically, most applications will require the controller status to be included in telemetry which means that controllers will have to implement interface Telemeterable (see section 8.3).

Similarly, they may have to implement other interfaces introduced to address other problems arising within the AOCS domain. These problems, however, have nothing to do with controller management and hence the decision as to which interfaces concrete controllers eventually have to implement is left to the architectural definition phase. It is only in this phase that abstract constructs defined at framelet level are integrated with each other to produce the kind of class map that is normally used to represent a framework in its final form.

The third characteristic feature of the design level is its emphasis on design patterns. The search for solutions at this level is essentially a search for design patterns to model the points of variations identified during the domain analysis. This search will often start from standard patterns but it will normally progress to the definition of new patterns that are optimized for the particular domain of interest. In the case study of section 8.4 the solution to the controller problem can be seen as an AOCS-specific design pattern. The pattern will then be instantiated during the framework architecture definition phase.

The example of section 8.4 also illustrates the primacy of design patterns and abstract interfaces over components as the fundamental building blocks of frameworks. The design solution proposed for the controller problem includes a core component, the controller manager, and possibly some default components representing commonly used types of concrete controllers. The definition of the concrete components is however straightforward as soon as the controller design pattern and its associated abstract interface have been defined and it is here that the real conceptual effort required to develop the framework lies.

8.10 From Design to Architecture

Framelets were variously defined as *self-standing units of design* or as *partitions of the space of design patterns and abstract interfaces*. Framelets simplify the framework design process by allowing the designer to tackle design problems in isolation from each other. In a framework context, the typical design problems concern the provision of adaptability and extensibility mechanisms for the software under development. Framelets gather together a small set of abstract interfaces and design patterns that solve a specific adaptability or extensibility problem in the framework domain.

In the case of the AOCS Framework, 13 framelets were identified. The criteria that were used to identify them were already briefly discussed in section 4.1.7. Essentially, one framelet was associated to each of the nine functionalities listed in section 8.2 and to each of the four infrastructure problems of section 8.7. By way of example, the present chapter gave an overview of two framelets – the *telemetry* and the *controller* framelets. As already noted, the level of detail at which the framelets were described in this chapter corresponds to the framelet concept definition phase. Its most distinctive feature is the mutual independence of the design solutions embodied in each framelet. This independence is useful at the beginning of the development cycle but, as the design progresses, it must give way to a more integrated view of the framework. Framelet integration is the crucial task in the transition from the design to the architectural phase. Although no general rules could be found in the AOCS Framework Project to guide the framework designers in this task, it is useful to sketch the way this integration process unfolded in the specific case of the AOCS Framework. Two main conceptual steps can be recognized.

In the first step, the functionality manager framelets were integrated. The functionality management framelets have a similar structure based on a manager component that holds a list of references to components (the functionality implementers) characterized by the same functionality interface. The functionality management framelets were integrated by deciding which functionality implementers inherit which interfaces. The resulting architecture is shown in figure 8.11 (using a semi-formal notation). The figure shows how the framelet boundaries become blurred in the transition to the architectural phase. The functionality implementers cannot be assigned to any framelet because they participate – through interface implementation – to several framelets at the same time.

Note that the same component can at the same time act as manager with respect to one functionality and as implementer with respect to a different functionality (in the figure, this is the case of the functionality manager 3). Thus, for instance, the controller manager in the AOCS Framework implements interface Telemeterable which makes it, from the point of view of the telemetry functionality, a functionality implementer.

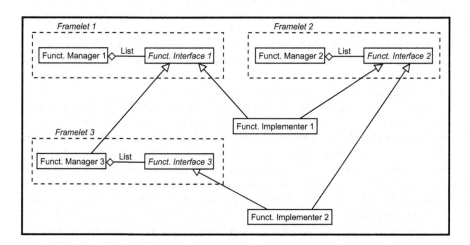

Fig. 8.11. Integration of the AOCS Framework functionality managers

In the second step in the integration of the AOCS framelets, the interaction between the functionality managers and the infrastructure problems was studied. In the case of the controller manager, for instance, it was realized that the same control loop may need to implement different control algorithms during different stages of the same mission. This means that controller components have operational mode. The framework provides a design pattern to implement mode-dependent behaviour (this is offered by the operational mode management framelet of chapter 12) and hence the solution to the controller management problem at architectural level is obtained by simultaneously instantiating *both* the controller and mode management design patterns. These design patterns originate in different framelets but, in the final form of the framework, are implemented by the same set of concrete components. This is another example of how, during the transition to architectural level, framelet borders were crisscrossed to the point where they lost their significance.

8.11 Related Work

To the knowledge of the author, the AOCS has never been the target of software frameworks or other attempts to promote domain-specific reuse. In the space field, the only known example of software framework is the OBOSS system [71] which proposes a reference architecture and a library of reusable components for the data handling part of satellite on-board software. From a functional point of view, there is some overlap with the AOCS Framework because the AOCS domain, as defined in the AOCS Framework Project, includes both telemetry and telecommand handling which are the central functionalities covered by OBOSS. The two frameworks are otherwise very different because OBOSS is underlain by an event-driven model which is dictated by the characteristics of data handling applications whereas the AOCS Framework assumes a cyclical application. Additionally, OBOSS is an object-based framework and relies heavily on Ada83 as an implementation medium whereas the AOCS Framework is component-based and object-oriented and, although the prototype was implemented in C++, could in principle be ported to other object-oriented languages.

If one further broadens the search for comparable work to avionics systems in general, the most prominent, and one of the earliest, examples of framework-like systems is ADAGE [10, 11]. Its target domain is avionics control systems where the term "control system" covers only the flow of data from position and attitude sensors through filters and compensators to the pilot and/or some actuation system. This is a much narrower understanding than adopted in the AOCS project whose domain includes functions like failure detection and recovery, telemetry and telecommand management, etc. On the other hand, the ADAGE project is more sophisticated than the AOCS project in being based on the principle of GenVoca generators [12] where the composition of components to build an application is performed automatically and checked for syntactical and semantic correctness. The formal rules that define the semantic correctness of a given composition of ADAGE components are the vehicle through which the reference architecture for the target domain is captured and made available to application developers.

It is seems unlikely that the ADAGE approach could be extended to the AOCS domain as defined here because of its greater complexity. GenVoca postulates that a software systems can be explained as a composition of primitive, plug-compatible and interchangeable components. This is not true for the AOCS domain as a whole. It may be true for its sub-domains (of which the control data flow modeled by ADAGE is one) and an AOCS application could perhaps be reconstructed as an intersection of several ADAGE-like systems.

Other frameworks described in the literature are nominally aimed at avionics system but in fact model mainly the interaction between the platform software and the operating system. Reference [1], for instance, proposes a reference architecture to encapsulate operating-system and scheduling aspects. Some of its solutions are similar to those found in the AOCS project. Its timer and data classes, for instance, have counterparts in the AOCS Framework. Other solutions are more sophisticated as is the case of its handling of scheduling issues which is far more flexible than for the AOCS Framework. The distinction made by [1] between component-level and platform-level classes recalls the discussion in section 8.6 of the AOCS Framework as a domain-specific extension of the operating system but the resemblance is very superficial. The platform-level classes should isolate the application from platform dependencies but they consist of abstract interfaces and not, as in the case of the AOCS Framework, of concrete components that are application-independent and that implement functions that are OS-like. The platform-level classes, in other words, are an *interface to* the operating system, they are not *part of* the operating system like the functionality managers of the AOCS Framework.

Reference [33] follows the same pattern and describes a framework to handle the flow of control in an avionics application. In an important difference from the AOCS Framework, the target application is assumed to be distributed and provisions are in place to shield application developers from knowledge of the physical architecture of the system. Distribution also plays a major role in the avionics reference architecture proposed by [102]. In both cases the distribution aspect is built upon the TAO infrastructure [45].

Reference [102] comes close in scope to the AOCS Framework because it explicitly models the data flow in the control loop and it includes provisions for reconfiguring the system in response to the detection of failures. However, it does not seem to model failure detection and recovery. Other functions that are part of the AOCS Framework – like manoeuvre management and telecommand and telemetry management – are also excluded.

Reference [89] presents a more general, non-avionics, framework for embedded monitoring and control systems which shares with the previous examples the emphasis on the distribution and data flow aspects. The computational paradigm it uses is more general, though, because there is no implicit assumption that the underlying system is cyclical as is normally the case in the avionics domain.

Finally, there are a number of recent reports of frameworks for non-avionics embedded systems but usually very few details are provided presumably owing to their company proprietary character. Typical examples can be found in [63, 31].

9 General Design Principles

The AOCS Framework was developed following the design process of chapter 6. The first phase in this process is the concept definition phase which in turn starts with a definition of the general design principles for the framework. The general design principles are the overall constraints on the framework design which define the policy to be adopted in regard of general design issues that affect the entire framework. Each of the design principles that were formulated for the AOCS Framework is discussed in a dedicated section in this chapter.

9.1 Boundary Conditions

Broadly speaking, the ultimate objective of the AOCS Framework project is to improve the re-usability – and hence lower the cost and enhance the reliability – of AOCS software. This objective is achieved by exploiting recent advances in software technology (the transition to the object-oriented paradigm) and in micro-processor technology (the space-qualification of SPARC processors). In general, technological advances can be used either to expand the functionality of a piece of software or to improve its quality. Historically, the emphasis has been on the former objective. This is why successive revolutions in software engineering – the transition to compiled languages, then to procedural languages, and still later to modular and object-oriented languages – have not had a major impact on software quality: programs are as error-prone and as poorly readable/reusable today as they were in the sixties! (See reference [21] for the classical argument in support of this contention).

The intention behind the development of the AOCS Framework is to concentrate on improving certain quality attributes, notably reusability and ease of extensibility. Hence, the framework is designed to cover only the present functionalities of AOCS systems. By constraining functionality to its present level, technological advances are leveraged to improve quality and reduce development costs.

9.2 An Object-Oriented Framework

The AOCS Framework is designed as an object-oriented software framework in the sense of section 3.1. Its design is based on pure object-orientation. It adds a further dimension – specific to frameworks – to the typical object-oriented design

A. Pasetti: Software Frameworks, LNCS 2231, pp. 131-140, 2002.

process because it places additional emphasis on identifying points of behaviour adaptation where the framework can be tailored to the need of a particular application. At the design level, adaptation is based on the use of design patterns. Wherever possible, the patterns of reference [44] are taken as a starting point and tailored to the needs of the AOCS domain. The development of a library of AOCS-specific design patterns is regarded as one of the important contributions of the AOCS Framework Project.

At the architectural and implementation level, adaptation is based on abstract coupling and dynamic binding and it is realized through inheritance or through object composition. The former mechanism is used for static behaviour adaptation, the latter for dynamic behaviour adaptation. In the present implementation, no effort is made to favour one mechanism over the other.

9.3 A Component-Based Framework

The AOCS Framework is component-based. At this stage, the term *component* is used in sense of section 3.1 to designate a binary unit of reuse that implements one or more interfaces and that can be configured for use in a specific application at run-time. This implies that modules defined by the framework can be reused without changes to the source code. It should also be noted that the AOCS Framework is implemented in C++. A component then becomes an object that is instantiated from a class that inherits from all the pure virtual classes representing the interfaces that it should expose. This means that, in this version of the framework, all components are objects and the two terms "component" and "object" are used somewhat interchangeably in the description of the framework.

The component-based character of the AOCS Framework paves the way for future developments where the framework will be implemented upon a proper component infrastructure à la DCOM or CORBA. In order to understand the benefits of doing this, consider again the example of telemetry management presented in the previous chapter. The proposed solution is built on a telemetry manager component that, when it is activated, calls the operations defined by interface Telemeterable on the components whose state should be included in telemetry. Telemetry management is a function that is usually performed by every subsystem on a satellite. If the subsystems are physically located on different processors, or even only in different memory partitions on the same processor, then each subsystem will require its own telemetry manager component. This is clearly wasteful. A more efficient architecture has a single telemetry manager that is able to see all telemeterable components as if they resided in the same address space. A component infrastructure provides precisely the mechanisms to make this possible. It is in view of this kind of generalization that the AOCS Framework was designed as a component-based framework. A concrete mechanism for achieving this kind of interoperability across processor barriers while remaining compatible with the real-time and criticality requirements of the AOCS software. is discussed in [22].

9.4 Delegation of Responsibility

If a certain task has to be performed by or upon an object, the responsibility for performing the task is as far as possible delegated to the object itself. Thus, for example, rather than having a procedure that writes the state of an object to telemetry, the object itself is endowed with the ability to write its won state to telemetry (i.e. it is given a method that, when called by an external telemetry manager, causes the object to write its own state to the telemetry stream). Delegation of responsibility to objects results in a "bottom heavy" implementation where many functions are implemented in the classes at the bottom of the class hierarchy. This is the price paid to ensure that the higher level classes that embody the AOCS structure are application-independent and can be reused without changes.

9.5 Multiple Implementation Inheritance

The AOCS Framework was implemented in C++ that allows multiple inheritance of implementation. As is well known, multiple implementation inheritance presents semantic problems that advise against its use [98]. No such problems, however, attach to multiple inheritance of interfaces (in the Java sense of the word). Hence the AOCS Framework Project adopted the following design restriction: only single inheritance of implementation is allowed, and multiple inheritance (or multiple implementation) of interfaces is allowed. This solution is borrowed from the Java language and it is both safe and well-tested.

9.6 External Interfaces

External interfaces – towards external devices and towards the processor – are encapsulated in components that, within the AOCS software, act as proxies for the external objects. This is an instance of the *proxy* pattern from reference [44].

9.7 Basic Classes

A *basic class* in the AOCS Framework is a class that serves as a base class to a large number of framework classes. Basic classes are introduced to allow uniform treatment of large numbers of framework components. Two basic classes are present in the AOCS Framework: RootObject and AocsObject. Class RootObject is used as the base for all concrete (non-interface) classes in the AOCS Framework and in the applications instantiated from it. This class defines two read-only attributes: the instance identifier and the class identifier. It is declared as follows:

RootObject
getInstanceIdentifier() : int getClassIdentifier() : int

Class `RootObject` ensures that all objects and all classes in the framework and in AOCS applications can be uniquely identified. The class identifier mechanism was not used in the framework prototype but could be useful in the future to provide a simple form of run-time type identification.

The `AocsObject` class is introduced as a base class for all non-trivial classes in the AOCS Framework and in the AOCS applications. Its purpose is to gather together and provide to its children classes some basic services that are likely to be required by all but the simplest objects in the framework. Objects that are directly or indirectly instantiated from `AocsObject` are called *AOCS Objects*. Class `AocsObject` makes the following services available to its children through protected methods:

– Recovery of current system time
– Reporting of failure conditions
– Reporting of configuration errors

Additionally, AOCS objects have the following properties that they inherit from abstract interfaces implemented by class `AocsObject`:

– Telemeterability (AOCS objects can be included in telemetry – see chapter 21)
– Resettability (AOCS objects are capable of resetting their internal state – see section 10.1)
– Configurability (AOCS objects are capable of checking that they are correctly configured – see section 10.1)

AOCS objects in an application are intended to be created during initialization and never to be destroyed. To ensure that this is the case, the `AocsObject` destructor generates an error if it is called. Note that this means that AOCS objects cannot be created on the stack or passed as method parameters. Only pointers to objects can be passed through method calls. Since AOCS objects are created before the application begins normal operation and are never destroyed, there is no danger of dangling pointers. A design rule in the AOCS Framework dictates that pointers can only be used on AOCS objects (namely on objects instantiated from classes derived from `AocsObject`). Adherence to this rule could in principle be checked by automatic inspection of the framework source code.

9.8 Time Management

The AOCS Framework assumes that two kinds of timing information are available: a clock time measurement from an AOCS clock component and a cycle number that indicates the number of AOCS cycles elapsed from some starting cycle. An AOCS clock is a component implementing the following interface:

AocsClock
getTime() : AocsTime getCycle() : int

The AOCS Framework assumes that all applications will make available one AOCS clock component. AOCS objects have access to timing information via the services offered by class `AocsObject`.

9.9 Language Selection

The framework design is intended to be language independent. The only assumption is that the implementation language should be powerful enough to support full object-orientation. This in particular implies support for the safe version of multiple inheritance used in the framework design.

The framework prototype is implemented in C++ but other language solutions are possible. Ada95 is still the recommended language for mission-critical real-time projects and ensuring compatibility with it would be correspondingly desirable. However, Ada95 does not directly support multiple inheritance. The effects of multiple inheritance can be achieved using the *multiple view* mechanism of [8] but this is awkward to use as it essentially represents a manual implementation of a construct that in languages like C++ and Java is automatically coded by the compiler. It may be suitable where usage of multiple inheritance is incidental but not in a system like the AOCS Framework where it plays a fundamental role.

There is one further point that deserves to be made on the contentious issue of language selection. The AOCS Framework – like much modern design practice – is moving towards component-based architectures. Components are reused as binary units and, ideally, should offer language interoperability. If this vision is realized, then design attention will shift away from the internal architecture of the component and towards the interface that it exposes. At that point, language issues will become much less relevant at system level. This situation already exists today for operating systems – arguably the most successful type of commercial components – which are normally bought as binary entities and where the issue of the language in which they were originally written is of no import.

9.10 Execution Time Predictability

AOCS systems must be *demonstrably schedulable*. An application is schedulable if it can meet all its deadlines under any operational conditions. The schedulability requirement derives from the hard real-time nature of the AOCS. Additionally, since AOCS systems are mission-critical, schedulability must be demonstrable in the sense that it must be possible to verify it statically by analysing the source code. A necessary condition for the demonstrable schedulability of a software application is that the execution time of any of its code segments be statically predictable. This requirement has a repercussion at framework level insofar as some of the code in an AOCS application is inherited from the framework (core and default components).

The AOCS Framework safeguards timing predictability primarily by avoiding constructs and operations that lead to non-predictable code. In practice, this leads to the avoidance of exception handling and dynamic memory allocation. The former presented no particular problem as the framework infrastructure offers a mechanism for reporting events that can be used in lieu of exceptions. The latter was instead rather difficult to achieve. Most existing frameworks are targeted at non-real-time applications and make extensive use of dynamic object creation and destruction. Avoiding such operations required some rethinking of typical design patterns used in frameworks and sometimes resulted in slightly awkward constructs.

As already mentioned, the AOCS Framework relies extensively on dynamic binding to model application adaptability. At first sight dynamic binding might seem to be incompatible with code execution predictability because of the impossibility of statically associating a definite piece of code to a particular method call. However, in embedded control systems there is no dynamic class loading and hence the number of methods that *might* be called is finite (and often small). It is therefore always possible to determine worst-case execution times and use this estimate in the schedulability analysis.

The heavy reliance of the framework on design patterns poses a more serious problem since some design patterns have a recursive structure that makes static timing analysis impossible. In such cases, timing predictability can only be retained by adding meta-information to the source code that specifies the maximum depth of the recursion. In order to make it easy to provide this information, the framework design documents clearly identify all recursive design patterns they introduce and explain how upper bounds on the depth of recursion can be inferred from their semantics. Where applicable, this point is addressed in the description of the design patterns for the AOCS domain introduced later in this book.

9.11 Scheduling

The AOCS Framework can be seen as a repository of reusable assets for AOCS applications. When demarcating its field of application, it was decided to leave scheduling aspects outside the framework (see section 5.3). This is because there already exist commercial operating systems that offer very effective reusable solutions to the problem of handling task scheduling. The AOCS Framework should therefore take advantage of these existing solutions (it should reuse them!) by building a bridge to them rather than attempting to duplicate them.

The framework assumes that the AOCS application is periodic but does not otherwise make other assumptions about the scheduling policy implemented by the AOCS applications to be instantiated from it. This decoupling of functional from timing aspects is achieved by introducing the *active component* abstraction. An active component is characterized by the following abstract interface:

Runnable
run(deadline : AocsTime) : void initialize() : void terminate() : void

Active components are also called *runnable components*. Their key method is run. A call to run will cause a component to perform all the actions associated to the current period. Method run takes one single parameter representing the maximum time available for the execution of the method. This parameter is not used in the current version of the framework but could be used in the future to implement some kind of cooperative scheduling policy. Consider for instance a telecommand manager that maintains a queue of telecommands due for execution. As will be seen in section 20.1, the telecommand manager is an active component. If it possessed an estimate of the time it takes to execute a telecommand (or at least had an upper bound for this quantity), it could check how many telecommands it can execute in the current AOCS period. In another example, consider a component that performs a background check on program integrity by computing the checksum on the code memory. Such a component would be likely to implement interface Runnable and if it knew how long it takes to process each memory location, it could determine how far it can go in carrying out the checksum check before having to return from method run to give other active components a chance to run.

Methods initialize and terminate are available to encapsulate actions that have to be performed when an active component is first scheduled and when it is descheduled.

The decoupling of the scheduling policy from the framework architecture is obtained by stipulating that method run can only be called from outside the framework (normally by an external scheduler or by an interrupt servicing routine) and that active components that are also core components do not make any assumptions about the frequency with which their method run is called, about the source of the call, or about the relative ordering in which they are called. Whenever they are activated by a call to method run, they take whatever action is appropriate at the time they are called without regard to how recently they were last called in the past or to how soon they may expect to be called again in the future. This insulates them – and hence the framework – from the AOCS scheduling policy.

As discussed in section 8.6, the framework conceptualizes an AOCS application as a bundle of functionality managers. As will be seen in the remainder of this document, the functionality managers are active components and must therefore implement interface Runnable. Functionality managers are obvious examples of core components since they are application-independent. Their run methods consequently implement not to make any assumptions about their calling pattern. Obviously, the components that implement the AOCS functionalities and that customize the functionality managers will rely on their methods being invoked according to some timing pattern but these components are application-specific and are defined and configured by the application developer at application level. Their dependency on a particular scheduling policy therefore does not carry over to the framework.

As an example, consider the controller functionality described in chapter 22 and already briefly outlined in section 8.4. A typical concrete controller component will implement a digital filter and therefore, in order to work properly, it will need to have its internal state propagated at regular intervals. Such a component will therefore have a built-in assumption about the frequency with which some of its methods must be called. No such assumption, however, applies to the controller manager component that sees the concrete controllers through an abstract interface whose operations do not imply any timing requirements.

The framework architecture therefore achieves independence from the scheduling policy of the AOCS application by confining scheduling assumptions to the application-specific components.

It should finally be noted that some scheduling policies require the implementation of synchronization mechanisms to coordinate access to shared resources. The framework regards any public method as potentially accessing a shared resource and therefore as potentially in need of an access synchronization mechanism. The type of mechanism to be used, however, is treated as an implementation rather than an architectural issue and is therefore neither provided nor dictated by the framework[14]. Note however that in the overwhelming majority of existing applications the AOCS software uses cyclic and non-preemptive scheduling and the issue of protection to shared resources does not even arise.

9.12 A Framelet-Based Framework

The AOCS Framework was developed as a framelet-based system. A total of 13 framelets were identified. They are listed with a brief description of their purpose and their contribution to reusability in table 13. The framelets were identified using standard heuristics as discussed in section 4.1.7. The remainder of the book devotes a chapter to each framelet. Their format is always the same. First, the design problem is outlined and the conventional solution as it is used in existing AOCS is briefly described. Then the main constructs offered by the framelet are presented at concept level. Design patterns usually have pride of place in recognition of the central role they play in the framework conceptualization proposed in

---

[14] This decision was partly dictated by the choice of C++ as the implementation language for the framework prototype. The framework is currently being ported to Java and this version will be thread-safe by default.

Table 13. Overview of framelets in AOCS Framework

Framelet Name	Framelet Description
System Management	Addresses problem of performing the same action on all components in the AOCS software. Enhances re-usability by decoupling the management of the system actions from their implementation.
Object Monitoring	Addresses problem of monitoring a component and its properties. Enhances re-usability by decoupling management of monitoring process from implementation of monitoring checks.
Intercomponent Communication	Addresses problem of managing data exchanges among components and defines a standard interface for data representing AOCS-specific quantities exchanged among these components. Enhances reusability in two ways: 1) it decouples production of data and events from their consumption; 2) allows uniform treatment of all data types.
Sequential Data Processing	Defines a pattern for the sequential processing of AOCS data. Enhances reusability by providing a standard interface for components that perform sequential processing on AOCS data and by allowing easy combination of data processing blocks.
Aocs Unit	Addresses problem of managing external AOCS units. Enhances reusability by decoupling managers and users of unit data from the units themselves.
Reconfiguration Management	Addresses problem of reconfiguring redundant functions. Enhances reusability by decoupling management of reconfigurations from the use of the redundant function and from the reconfiguration algorithm.
Operational Mode Management	Allows components to be made operational mode-dependent. Enhances reusability by separating the implementation of mode-specific algorithms from mode switching logic.
Manoeuvre Management	Addresses problem of managing manoeuvres. Enhances reusability by separating management of the manoeuvres from their implementation.
Failure Detection Management	Addresses problem of managing failure detection checks. Enhances reusability by decoupling management of failure detection function from implementation of failure detection checks.
Failure Recovery Management	Addresses problem of managing failure recovery. Enhances reusability by decoupling management of failure recovery function from implementation of failure recovery responses.
Telecommand Management	Addresses problem of managing telecommands. Enhances reusability by decoupling management of telecommands from their implementation.
Telemetry Management	Addresses problem of managing collection of telemetry data. Enhances reusability by decoupling management of telemetry from format and content of telemetry data.
Controller Management	Addresses problem of managing closed-loop controllers. Enhances reusability by decoupling management of controllers from implementation of control algorithms.

this book. Where appropriate, implementation cases are also presented. Their choice was dictated by didactic considerations – the desire to show how the main framework constructs are built and used. It is not necessarily representative of the type of implementation cases that would be used in a real development environment where the emphasis is more on checking the framework design and its ability to adapt to application requirements. The level to which the implementation cases are worked out is of course that typical of the concept definition phase and much is therefore left open. The framelet chapters close with a brief discussion of the reusability of the framelet constructs both within the AOCS domain and in other domains. A table is also given that lists the framelet constructs defined or identified in the chapter[15]. As a rule, the design patterns are fully defined whereas all other constructs – components, abstract interfaces and hot-spots – are either just identified as necessary or defined in an incomplete manner. Their full definition would normally be done in the architectural definition phase which, however, is not covered in this book.

---

[15] See section 6.4 for the distinction between *defining* a construct and *identifying* it.

10 The System Management Framelet

An AOCS software instantiated from the AOCS Framework will be constituted of a set of interacting components. A *system management function* is a function that is performed systematically on all AOCS components present in the AOCS software at a given time. Typical system management functions that are found in many AOCS applications are:

- *System Reset*: A system reset causes the internal state of all AOCS components to be reset to a default state.
- *System Configuration Check*: A configuration check causes the configuration of all AOCS components to be checked.

This type of system-wide blanket operations are typically necessary during system start-up or after a system re-initialization following a complete software reset. Existing AOCS applications do not explicitly recognize system management functions. This is because like most embedded control applications, AOCS applications, although nominally object-based, use the object metaphor only for purposes of information hiding. No effort is usually made to treat the individual objects as self-standing entities capable of offering services to their clients. The latter is instead the view taken in the AOCS Framework which sees the AOCS software as truly component-based. This view gives rise to the need for a mechanism to operate on all components at the same time in a simple manner. This is the design problem that is addressed by the system management framelet. The framelet provides a design pattern to handle generic system management operations and a core component that performs the system reset and configuration checks functions in an application-independent manner.

10.1 The System Management Design Pattern

The system management design pattern is introduced to address the problem of systematically performing the same set of operations on a target set of components. The system management design pattern is obtained by specializing the manager meta-pattern. Its structure is shown in the UML diagram of figure 10.1.

The functionality interface encapsulates the operations to be performed on the target set of components. This interface is to be implemented by all components in the target set. Additionally, all components in the target set are to be derived from a single base class whose constructor registers with the functionality manager.

A. Pasetti: Software Frameworks, LNCS 2231, pp. 141-145, 2002.
© Springer-Verlag Berlin Heidelberg 2002

Thus, the functionality manager is automatically provided with a list of all the components in the target set. Deregistration is not required because no dynamic object destruction can take place (see section 9.7).

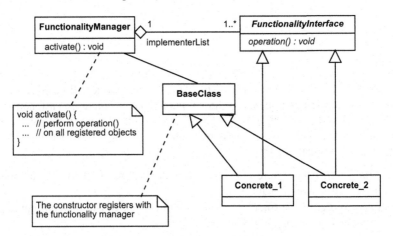

Fig. 10.1. Class structure of the system management design pattern

In the AOCS Framework, the system manager design pattern is instantiated twice, once for the system reset function and once for the system configuration check function. A single component called the *system manager* acts as functionality manager for both the system reset and system configuration check functions. The base class is identified with the `AocsObject` class (see section 9.7) which means that all components in an AOCS application can be acted upon by the system manager. The abstract interfaces associated to the system reset and system configuration checks functions are called `Resettable` and `Configurable`, respectively, and are discussed in the next two subsections. These interfaces are implemented directly by class `AocsObject` and are therefore inherited by all framework components. The class diagram of the system management design pattern thus instantiated is shown in figure 10.2.

10.2 The System Reset Function

The system reset function is defined by the `Resettable` interface:

Its key method is `reset`. A call to method `reset` causes a component's internal state to be brought back to where it was when the AOCS application began

normal operation but after it had been configured during the application initialization. A call to method `reset` therefore resets the attributes of a component that are updated as part of the component's performing its allotted task but leaves unaltered the attributes that are normally set during the application initialization to configure it. As an example, consider a component implementing a digital integrator. Its resettable state consists of the variables that are updated every time the integrator is triggered. A call to method `reset` on the integrator brings it back to the state it had after the application had been fully initialized and was ready to start normal operation.

In the AOCS Framework a reset operation can be performed at two levels. At *component's level* an individual component is reset by calling its `reset` method. At *system level*, a reset is performed by asking the system manager to reset all the components in the AOCS application.

In the AOCS Framework a system reset is distinguished from a system reboot. The system reboot will cause the AOCS application to be reloaded and restarted. A system reset will simply reset the state of all components. The system reset therefore does not destroy the configuration of the AOCS application. The system reset function is provided mainly for failure handling as it represents a less disruptive way of handling failures than a complete system reboot. The `reset` method is a hot-spot (the *resettable hot-spot*) because it has to be defined for each specific component in an AOCS application.

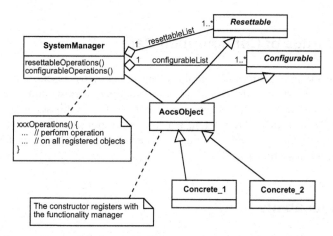

Fig. 10.2. Class structure of the instantiated system management design pattern

10.3 The System Configuration Check Function

AOCS components are configured during initialization for use in a specific application. Configuration will typically imply loading plug-in components and setting the value of configuration parameters. In order to verify that an application is correctly configured, configurable components expose a method that reports whether

they are configured or not. This functionality is defined by interface Configurable:

Configurable
resetConfiguration() : void
isConfigured() : bool

The key method is isConfigured that returns false if the component is not
properly configured. Often, this will simply mean that there are undefined references or that configuration parameters have illegal values. The system manager can
call isConfigured on all components in an application and check whether any
of them reports false. In this way, the configuration of the entire application can
be verified at any time.

Method resetConfiguration clears all configuration information in a
component and brings it back to the state in which it was at the time it was created. Note the difference with method reset that brings the state of a component
to where it was at the end of the configuration process.

The isConfigured and resetConfiguration methods constitute a
hot-spot (the *configurable hot-spot*) because they have to be defined for each specific component in an AOCS application.

10.4 Storage of Configuration Data

There is some information that should be preserved across resets. This in particular concerns the state of the *reconfiguration managers* (see section 16.2). This information should be preserved because it defines the health status of redundant
functionalities or units. The system manager is responsible for maintaining and
restoring the state of the reconfiguration managers. This issue is discussed in
greater detail in section 16.2.3.

10.5 Reusability

Table 14 summarizes the constructs exported by the framelet that have been defined or identified in this chapter.

Reusability of the framelet constructs within the AOCS domain is achieved by
decoupling the management of the reset and configuration checks from their implementation. The system manager only sees an abstract interface and is thus protected from details of how these functionalities are implemented. The system manager is therefore application-independent and is a core component because it
would be found in all AOCS applications instantiated from the framework.

Neither the reset nor the configuration check operations are specific to the AOCS. Hence, this framelet can be reused in other domains where it is desired to perform these operations on all (or a large number of) components in an application. Reusability in such domains can span al three reuse levels if the system management functions are the same as in the AOCS domain as in this case both the `resettable` and `configurable` abstract interfaces and the system manager components can be reused without changes. If the system-wide operations are different, then reuse is in principle possible only at the design level by reusing the system management design pattern. Extension of the framelet to cover other system-wide operations could however be easy because it would entail introducing abstract interface to model the additional system-wide operations and subclassing the system manager class to add to it operations to handle them.

Table 14. Overview of system management framelet

Design Patterns	
System Management	Allow the same set of operations to be systematically performed on all components in an AOCS application.
Abstract Interfaces	
`Resettable`	Characterize a component whose state can be brought back to the value it had at the end of the application initialization and at the beginning of normal operation.
`Configurable`	Characterize a component that can check whether it is configured.
Core Components	
`SystemManager`	Offer services to perform a system reset and a configuration check on all components in an AOCS application.
Hot-Spots	
Resettable	Applications must provide an implementation for interface `Resettable` to reset a component's state.
Configurable	Applications must provide an implementation for interface `Configurable` to clear a component's configuration and to check whether it is configured.

11 The Object Monitoring Framelet[16]

An AOCS needs to perform a certain amount of monitoring of internal variables for purposes of failure detection and to allow components to synchronize their behaviour. Which variables should be monitored and with respect to which behaviour will of course vary across applications. The monitoring mechanism itself should however be an application invariant in order to ensure that the same function – the monitoring of a variable – is performed in the same manner both within the same application and in different applications. This generic monitoring mechanism must therefore be defined by the AOCS Framework. To solve this problem, the framework offers two design patterns which together with their associated hot-spots form the core of the *object monitoring framelet*. The monitoring mechanism that it proposes is loosely based on the JavaBeans property model [36]. It extends it by allowing monitoring to be performed with respect to changes of a specified kind. In the JavaBeans architecture, a "change" is defined as any variation in the value of the observed property. This would be too broad for AOCS applications where monitors are not interested in any change in the variables they monitor but are instead interested in being alerted only if the monitored variables fail to follow pre-specified behaviours (such as remaining below a threshold, or not exhibiting "jumps", etc).

The object monitoring design patterns are based on two ancillary concepts: property objects and change objects. This chapter begins by introducing these two concepts and then proceeds to present the framelet design patterns.

11.1 Properties and Property Objects

A *property* is an attribute of a component that describes one aspect of its behaviour or of its internal state and that is accessible to external components. Properties are obviously useful for failure detection purposes: the mode of change of a component's properties may indicate that a fault has arisen. Components also monitor each other's properties when they want to coordinate their behaviour. Monitoring actions are therefore performed upon properties.

The property model adopted by the AOCS Framework is derived from the JavaBeans component architecture [36]. Components expose their properties

---

[16] The framelet should really be called "component monitoring framelet". Unfortunately, its name was selected before the framework was given the a component-based character.

A. Pasetti: Software Frameworks, LNCS 2231, pp. 147-157, 2002.

through `get` and `set` methods. A property of name `<PropertyName>` and type `<PropertyType>` has getter and setter methods that conform to the following signature and naming pattern:

```
<PropertyType> get<PropertyName>();
void set<PropertyName>(<PropertyType> value);
```

Components that wish to directly monitor a component's property call its `get`
method. Components that wish to set the property call its `set` method. Some properties are read-only and do not provide a set method. Properties of boolean type
may have an `is<PropertyName>` method to check their value

It is often useful to be able to treat properties as entities of the same syntactic
type. Consider for instance the case of a failure detection manager. This component will need to keep a list of properties that must be monitored regularly. It
would be convenient if an array of properties could be defined and if some kind of
standard operations could be performed on all properties. In order to make this
possible, *property objects* are defined. Property objects encapsulate a property and
give it a standard interface and type. They are conceptually defined by the following interface:

Property
get() : float

The `get` method returns the value of the property. This value is (or can always
be cast to) a float. In a C++ implementation a property object is essentially a read-
only wrapper for a pointer.

To see the advantages of having property objects, consider the case of a failure
detection manager that needs to perform systematic checks on several unrelated
properties. It can do so by maintaining and inspecting a list of properties as follows:

```
Property* properyList[20];
float value;
. . .
for (int k=1; k<20; k++)
{   value = property[k]->get();
    . . . // process 'value'
}
```

Later in the chapter, it is shown how the processing of the property value can
be made in a generic fashion.

11.2 Change Objects

The term *monitoring* in the AOCS Framework refers to the observation of changes
in property values. At the most basic level, monitors are interested in *any* change

in the monitored property. More frequently, however, they are only interested in *certain types* of changes. For instance, a failure detection manager may be interested in knowing when the output of a sun sensor exceeds certain pre-specified boundaries, as this might indicate that the spacecraft is going out of control or that the sensor itself is faulty. If the failure detection manager were notified every time the sun sensor output changes, it would have to implement a sensor-specific filtering mechanism to identify the out-of-range changes that might signal a failure. In order to avoid burdening monitors with the need to implement highly specific (and hence non-reusable) filters, the concept of change is encapsulated in objects which will be called *change objects*. Thus, a change object is an object that encapsulates a specific time profile for a property value. The change is said to have occurred when the property value has violated the time profile.

A change object may encapsulate any time profile. Typical change objects that are useful in an AOCS (and that are therefore provided as default components by the framework) include:

– *Simple change*: the change occurs when the monitored property changes its value.

   Example: the mode manager of an component registers with the mission mode manager to be notified whenever the mission mode changes (see chapter 12 for the background on mode management).

– *Out-of-range change*: the change occurs whenever the monitored property moves outside a pre-defined range.

   Example: when a sensor output drifts outside a range of values representing physically possible values, a fault is likely to have arisen. Hence, the failure detection manager registers with the components representing sensors to be notified when their outputs leave the permitted range.

– *Delta change*: the change occurs whenever the monitored property changes by more than a pre-defined delta value.

   Example: when the output of a sensor suddenly "jumps" by more than a certain threshold, a fault is likely to have arisen. Hence, the failure detection manager registers with the components representing sensors to be notified when their outputs change too abruptly.

– *Filtered delta change*: monitors are notified whenever the filtered value of the monitored property changes by more than a pre-defined threshold.

   Example: the failure detection manager of the previous example may wish to ignore "spikes" in sensor outputs that might be due to freak conditions not indicative of a real fault. In that case, it registers with the sensor components to be informed whether its output, after filtering for spikes, exhibits a suspicious change.

Other types of changes may be defined on an *ad hoc* basis. The object monitoring framelet introduces the following interface to characterize a change object:

ChangeObject
checkValue(value : float) : bool

Instances of this type are responsible for carrying out the check on a property value and deciding whether a specified change criterion is met. It is assumed that the property value is always a float or can be meaningfully converted to a `float`.

Method `checkValue` takes as argument the current value of the property being monitored and it returns `true` if this value represents a violation of the change profile encapsulated by the change object. Note that change objects will in general encapsulate profiles that extend over time and hence change objects may need to keep track of past values of a property.

Change objects give rise to a framework hot-spot (the *change object hot-spot*) since applications must supply the definition of the change profile in which they are interested. The framework offers several default components implementing common change profiles such as those listed at the beginning of this section.

11.3 The Monitoring Design Patterns

The term *monitoring* or *monitoring action* refers to the observation of a change over time in the value of a property. The component performing the monitoring action is called the *monitor*. The property being monitored is called the *monitored property*. Two design patterns are offered to address the monitoring problem and are described in the next two subsections.

11.3.1 The Direct Monitoring Design Pattern

The direct monitoring design pattern is introduced to address the problem of granting access to an internal property to a monitoring component. This pattern is very simple as it prescribes that the monitor directly accesses the monitored property through its getter methods. Thus, for instance, a component that is interested in monitoring property <property> belonging to component <component> will periodically access it by calling method get<property> on <component>. The pattern is illustrated in the class diagram of figure 11.1.

If the monitor is interested in detecting changes of a certain type, recourse can be made to a change object. Consider for instance the case of a monitored component of type SunSensor and suppose that the monitor is interested in detecting whether its X output is out of range. This can be done with the following statements:

```
output = aSunSensor->getXoutput();
if ( aOutOfRangeChange->checkValue(output) )
    . . .    // X output is out of range!
```

Object aOutOfRangeChange is a change object that encapsulates an out-of-range profile (see section 11.5 for a sample definition of its class).

In another example, consider a failure detection manager that keeps a list of property objects that must be checked periodically and systematically. To each property, a change object is associated that specifies the type of change that needs to be detected:

```
Property* propertyList[20];
ChangeObject* changeList[20];
float value;
   . . .

for (int k=1; k<20; k++)
{   value = propertyList[k]->getValue();
    if ( changeList[20]->checkValue(value) )
        . . . // the change has occured!
}
```

This example generalizes the example presented at the end of section 11.1.

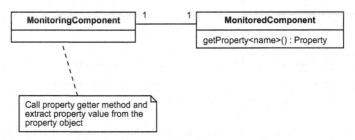

Fig. 11.1. Class diagram of the direct monitoring design pattern

11.3.2 The Monitoring through Change Notification Design Pattern

This design pattern is introduced to address the problem of monitoring for a component that wishes to be automatically notified when a change of a certain type occurs in a specific property in which it is interested. The JavaBeans component architecture [36] implements a simple mechanism of monitoring through change notification with the bound property mechanism: property monitors can register with the owner of a property to be notified whenever the value of that property changes. The JavaBeans mechanism is rather limited because it does not discriminate between different types of change: any change results in a notification. In the case of the AOCS, the change notification design pattern uses the change objects to ensure that notification of a monitor only takes place when a change of a prespecified type has occurred. Its UML diagram is shown in figure 11.2.

The property owner must allow monitors to register and unregister their interest in a property. It does so by exposing methods with signatures like:

```
void add<Property>Monitor(PropertyMonitor* monitor,
                          ChangeObject* changeObject);
void remove<Property>Monitor(PropertyMonitor* monitor);
```

When a monitor `monitor` calls `add<Property>Monitor` it notifies the property owner that it is interested in property `<Property>`. The second parameter in the method call indicates the type of change in which the monitor is interested. When a monitor `monitor` calls `remove<Property>Monitor` it notifies the property owner that it is no longer interested in property `<property>`.

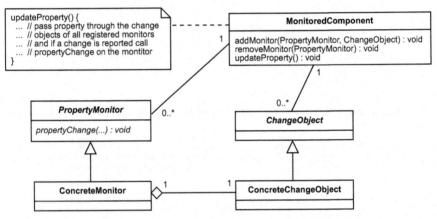

Fig. 11.2. Class diagram of monitoring through change notification design pattern

The property owner maintains a list of registered monitors together with their change objects and every time the property is changed, the new value is passed through its corresponding `checkValue` method. Monitors are notified through a call to method `propertyChange`. They must therefore implement the following interface:

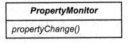

The arguments of `propertyChange` serve to identify the property that has undergone the change and its owner. The code for updating the value of property `property` would then look like this:

```
property := newValue;
for (all currently registered monitors) do {
    if (changeObject->checkValue(newValue)
                                        // change occurred!
        monitor->propertyChange(…);     // notify monitor
}
```

Note that, as in the JavaBeans architecture, the monitors are notified *after* the change has taken place. In keeping with the notation used by the JavaBeans architecture, properties that offer a notification mechanism are called *bound properties*.

The action to be taken in response to the detection of a change in a monitored property is application-specific and therefore method `propertyChange` defines a framework hot-spot (the *property change hot-spot*).

11.4 Implementation Case Example – 1

Consider the implementation case described in table 15. This section shows how it can be worked out during the framework design definition stage. Its objective is to perform a monitoring check on a property exposed by some component to verify that its value remains constant. The procedure for working out this implementation case is as follows. The component that owns the property to be monitored is endowed with a getter method following the naming pattern given in section 11.1. A change object is then created that embodies the behaviour with respect to which the monitoring is to be performed. The change object is used to perform the monitoring action. These three steps are now examined in greater detail.

The component that owns the property could be defined as follows:

```
class Component {

    float prop;
    . . .
    float getProp() {     // property getter method
        return prop;
    }
}
```

A change object that checks whether a variable remains constant can be defined as follows:

```
Class SimpleChange: public ChangeObject {

    float oldValue;

    bool checkValue(float newValue) {
        if (newValue != oldValue) {
            return true;
        }
        oldValue := newValue;
        return false;
    }
}
```

Here method `checkValue` checks whether the new value is different from the old one (which it remembers from the previous activation) and, if it is, it returns `true` to signal that the change has occurred. This kind of change is so common that its corresponding change object should be provided as a default component by the framework.

The last step in working out the implementation case is to show how the actual monitoring action can be performed:

```
Component* component;
ChangeObject* changeObject = new SimpleChange();
float monitoredValue;
. . .
monitoredValue = component->getProp();
if (changeObject->checkValue(monitoredValue))
    . . .   // change has occurred
else
    . . .   // monitored property is still unchanged
```

The important point to note is that the monitoring check is performed in a manner that is independent of the specific time profile against which the target property is being verified.

Table 15. First implementation case example for change object

Objective	Check that a certain property does not change its value.
Description	Components in an AOCS application must sometimes monitor the value of certain key variables to ensure that they behave as expected. This IC shows how a check can be perform to verify that a specified variable retain the same constant value.
Framelet	Object Monitoring Framelet
Constructs	`ChangeObject` interface
Hot-Spots	Change object hot-spot
Related ICs	None
Implementation	- Build a change object to encapsulate a constant time profile - Pass the property through the change object filter

11.5 Implementation Case Example – 2

Consider the implementation case described in table 16. Its objective is to perform a monitoring check on a property exposed by some component to verify that its value remains within a prespecified range. This implementation case is an extension of the implementation case of the previous section. The main innovation is that now a different implementation of the `ChangeObject` interface is required that verifies that the target variable remains within a certain range rather than constant. This could be built as follows:

```
Class OutOfRangeChange: public ChangeObject {

    float maxValue, minValue;
    bool checkValue(float value) {
        if ( (value>maxValue) || (value<minValue) ) {
            return true;
        }
        return false;
    }
}
```

In this case, `checkValue` returns `true` if the value it inspects is outside a predefined range. The monitoring code then becomes:

```
Component* component;
ChangeObject* changeObject = new OutOfRangeChange();
float monitoredValue;
. . .
monitoredValue = component->getProp();
if (changeObject->checkValue(monitoredValue))
    . . .    // change has occurred
else
    . . .    // monitored property is still unchanged
```

Note that the monitoring code is still the same as in the previous implementation case. This demonstrates that the proposed mechanism for monitoring actions decouples the execution of the monitoring action from the profile against which the monitoring is to be performed.

Table 16. Second implementation case example for change object

Objective	Check that a certain property remains within a specified range.
Description	Components in an AOCS application monitor the value of key variables to ensure that they behave as expected. This IC shows how a check can be performed to verify that a variable remains within a specified range.
Framelet	Object Monitoring Framelet
Constructs	`ChangeObject` interface
Hot-Spots	Change object hot-spot
Related ICs	This IC extends the IC of section 11.4.
Implementation	- Build a change object to encapsulate a fixed range profile - Pass the property through the change object filter

11.6 Alternative Solutions

In the sample implementation proposed here, notification of a change causes a generic method `propertyChange` to be invoked on the property monitor. Since the same component might be monitoring several properties or states, monitors have to discriminate among several types of change notifications. Sometimes, it may be more efficient to use some kind of adaptation mechanism that directly routes the property change event to its handler. This would be similar to the *event adapter* mechanism of the JavaBeans architecture. In the present version of the framework, the identification of the change that triggered a call to method `propertyChange` is done though its parameters. Use of an event adapter mechanism might be introduced in future versions if it is found that monitors typically have to monitor several change sources.

The JavaBeans architecture also envisages *constrained properties* whose change can be vetoed by the property monitor. It is unclear whether constrained properties are needed in the AOCS. They offer flexibility in coordinating state changes between components but, by the same token, they also increase the coupling between the components. Since the objective is to minimize inter-component coupling, the present version of the framework does not use constrained properties but this decision may have to be reviewed in later versions.

11.7 Reusability

The encapsulation of properties in property objects furthers software reusability because it helps decouple the task of *managing* the properties from that of *processing their values*. The encapsulation of change profiles in change objects serves the same purpose because it helps decouple the task of *managing* the checks on property changes from that of performing the checks. These two types of decoupling will in particular be crucial to building a reusable failure detection manager as will be seen in chapter 18. More generally, they allow generic – and hence reusable – monitoring code to be written. The separation between the monitors and the monitoring checks that they perform is formalized in the framelet design patterns.

The concept of change object can be used "as is" outside the AOCS domain wherever there is a need to encapsulate variations in a component's attribute more complex than a straightforward change in its value. The monitoring design patterns can also be used in non-AOCS contexts. It would be useful as a more sophisticated alternative to the JavaBeans type of monitoring providing a concept of "change" that covers deviations from a pre-specified behaviour.

Table 17 summarizes the constructs exported by the framelet that have been defined or identified in this chapter.

Table 17. Overview of object monitoring framelet

Design Patterns	
Direct Property Monitoring	Monitor a property by directly accessing its value through the getter methods exposed by the property owner.
Change Notification Monitoring	A monitor arranges to be notified whenever a certain variable changes its behaviour in a prespecified manner.
Abstract Classes	
`Property`	Encapsulate a property by offering a uniform interface through which to access its value.
`ChangeObject`	Characterize a component that encapsulates a change profile for a variable.
`PropertyMonitor`	Characterize a component that can ask to be notified in case a predefined change in a monitored variable takes place.
Default Components	
Concrete change object classes	Encapsulation of commonly used change profiles like simple change, out of range change, delta change.
Hot-Spots	
Change Object	Applications must provide the change objects encapsulating the particular time profiles in which they are interested.
Property Change	Applications must define the actions to be taken when a property subjected to monitoring through change notification has undergone a change. This is done by providing implementations for interface `PropertyMonitor`.

12 The Operational Mode Management Framelet

Current AOCS systems are based on the concept of operational mode which allows the behaviour of the software to be dynamically adapted to changing operational conditions. A software framework for the AOCS needs to supply a general mechanism to handle operational mode changes. This is the design problem that the operational mode framelet sets out to solve.

In current systems, the operational mode is an attribute of the AOCS software as a whole. In addressing the problem of operational mode management from the perspective of a component-based framework, the first basic design choice is between keeping a single operational mode for the whole AOCS system or making operational mode a property of individual components. The former approach requires a "mode manager" component which is responsible for maintaining the system mode and for ensuring that each component behaves in a manner that is consistent with the current mode. This component would act as a centralized coordinator of component behaviour. As such, it would require a rather detailed understanding of how each component is constructed and of what its internal state is. This approach would therefore weaken behaviour encapsulation thus running counter to the spirit of the AOCS Framework. In an alternative variant, the single mode manager acts as a client for other components that periodically sample it to ask what the current mode is and use this information to tune their behaviour. The drawback of this solution is that different components may need different sets of modes and a centralized mode manager needs to keep track of all of them which again weakens data encapsulation and component decoupling.

For these reasons, this framelet make operational mode a property of each component. Components then become responsible for updating their own operational mode in response to changes in the environment around them. A consequence of this choice is that components need to keep track of changes in their environment by monitoring the properties of other components. At implementation level, this can be done by using the monitoring mechanisms offered by the monitoring framelet of chapter 11. Mode-dependent components register with the properties of interest to them and ask to be notified when their values change in a predefined manner. They respond to a change in a property value by updating their operational mode.

Note that a monitored property could also be the operational mode of another component. Hence, the traditional architecture with an AOCS-wide operational mode could be implemented by having each component implement the same set of modes and by having components change their own mode to follow changes initiated by a centralized mode manager component.

A. Pasetti: Software Frameworks, LNCS 2231, pp. 159-165, 2002.
© Springer-Verlag Berlin Heidelberg 2002

As an example of the decentralized mode concept proposed here, consider an attitude controller component. The component is responsible for implementing the attitude control algorithm. A controller component could have two modes: low accuracy control, and high accuracy control. The switch between the two modes could be done, for instance, on the basis of the size of the attitude error with the high-accuracy mode being used when the attitude error is very small and the low-accuracy mode being used when the attitude error is large. The controller component would then register its interest in the attitude error and would ask to be notified when this error crosses a certain threshold. The notification that the attitude error has crossed the selected threshold would trigger a mode change.

Mode management in the AOCS Framework is built around a major design pattern (the *mode management design pattern* of the next section) and gives rise to major hot-spots where application developers must specify the mode architecture of their components and the associated mode switching logic. An interface to encapsulate the actions to be performed at the transition between modes is also part of the offerings of this framelet.

12.1 The Mode Management Design Pattern

This design pattern is introduced to address the problem of endowing components with mode-dependent behaviour. The pattern separates the implementation of the mode-dependent behaviour from the implementation of the logic required to decide mode switches.

A mode-dependent component is a component that, depending on operational conditions, needs to select a particular *implementation* of one or more *strategies*. Figure 12.1 illustrates the case of a component with two strategies. The mode dependent component uses a mode manager to supply to it the implementations of the strategies that are appropriate for a given operational situation. The two strategies are seen as instances of classes `Strategy_1` and `Strategy_2`. These could be concrete classes, or base abstract classes, or abstract interfaces. The mode manager is characterized by the ability to provide instances of these two classes to the mode-dependent component. It must therefore implement an abstract interface like `ModeManager` in the figure that defines two getter methods for the two strategy implementations and two loader methods to associate the strategy implementations to the various modes managed by the mode manager. Concrete implementations of the `ModeManager` interface provide the logic to switch between modes. This is implemented in a method like `updateMode` that sets the current mode and therefore indirectly determines which implementers of the strategies are returned by calls to the getter methods.

The mode manager encapsulates the logic to decide which implementation of each strategy is appropriate at any given time. When the mode dependent component needs a strategy implementation, it uses the strategy getter methods offered by the mode manager to obtain one. From the point of view of the mode-dependent component, the strategy implementations are always seen as instances of type Strategy_1 and Strategy_2.

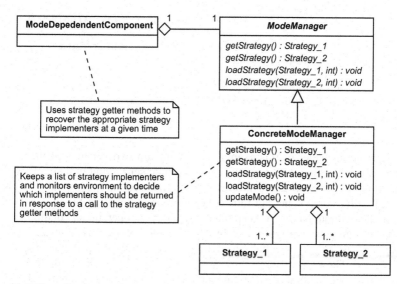

Fig. 12.1. Class diagram of mode manager design pattern

As an example consider again an attitude controller component[17]. This component is responsible for implementing the attitude control algorithm and could have two modes: low accuracy control, and high accuracy control. In low accuracy mode, attitude information is provided by a low-accuracy sensor, for instance a coarse sun sensor (CSS), and is processed by a low-accuracy algorithm, for instance a PID. In high-accuracy mode instead, a high accuracy sensor, for instance a fine sun sensor (FSS), and a high accuracy control algorithm, for instance a PID with Kalman filtering, are used. In this case, the controller component (the mode-dependent component) has two strategies (the sensor and the control algorithm) and each strategy has two implementations (the CSS and the FSS for the sensor strategy, and the PID and PID+KF for the control algorithm strategy). Thus, a skeleton code for an AttitudeController component could be as follows:

```
class AttitudeController {

    AttitudeSensor* sensor;
    AttitudeControlAlgorithm* controlAlgorithm;
    AttitudeControllerModeManager* modeManager;
    . . .
    void computeControlAction() {
```

---

[17] The mechanism prescribed by the framework to perform attitude control is different from that described in this example. The example is used for convenience only.

```
    // Retrieve strategy implementers
    sensor = modeManager->getAttitudeSensor();
    controlAlgorithm =
                    modeManager->getControlAlgorithm ();

    // Use first strategy to acquire sensor data
    sensor->acquireData();

     // Use second strategy to process the sensor data
    controlAlgorithm->processSensorData();
  }

}
```

The mode manager of this example will maintain two versions of the attitude control algorithms and two versions of the attitude sensor and will supply the appropriate one to the attitude controller algorithm based on operational conditions. The controller component can thus concentrate on computing the processing the sensor data without having to keep track of changes in operational conditions.

This design pattern gives rise to two framework hot-spots. The *mode manager hot-spot* is used by applications to define the application-dependent mode switching logic. The *mode implementer hot-spot* is used by applications to load the application-specific strategy implementations. In the previous example, the mode manager hot-spot is filled by creating the `AttitudeControllerModeManager` class whereas the mode implementer hot-spot is filled by defining the attitude sensors and control algorithms associated to each mode. For an example of how a mode manager can be constructed, see section 19.4.

The mode management design pattern is *instantiated* within the framework because it is used to model the mode-dependent behaviour of some of the framework core components (e.g. the failure detection manager, the failure recovery manager, the telemetry manager, etc). It is also *exported* by the framelet because it is made available to application developers to be instantiated for application-specific mode-dependent components that need to implement mode-dependent behaviours.

The mode management design pattern separates the implementation of the mode-dependent behaviour from the implementation of the mode-switching logic. One notable advantage of this approach is that mode-dependent components appear to their clients as modeless (since the mode switching logic is located elsewhere – in the mode manager component). This enhances ease of use because, in general, modeless components tend to be easier to use than mode-dependent components [52].

There is an obvious relationship between mode management and state machines. A mode manager can be seen as a state machine with as many states as there are operational modes and with a set of implementers associated to each state. This framelet adopts a mode management view and terminology simply because this seems to be the most common within the AOCS domain but it clearly does not prevent (and neither does it dictate) implementation of a mode manager as a state machine.

12.2 Mode Change Actions

Often, a mode switch by a component must be accompanied by some special actions related to the exit from the old mode and/or the entry into the new one. Consider for instance an attitude controller component that manages several control algorithms corresponding to different operational modes. When a new mode is entered, its associated control algorithm should be reset to remove the memory of previous activations. In another example, consider a unit trigger (namely a component introduced in section 15.3 to control the exchange of data with external units). To each of its modes a certain set of units is associated. When the unit trigger leaves a mode, it might have to switch off units that are no longer needed in the new mode and, conversely, when it enters the new mode it should turn on and initialize the newly needed units.

Actions that are associated to a mode transition are called *mode change actions*. The AOCS Framework offers two mechanisms for their implementations. The most straightforward approach is simply to hard-code the mode change actions into the concrete mode managers. Thus, for instance, in the case of the attitude controller example, when the mode manager finds that it has to perform a change of mode, then it executes any actions that are associated to the mode change. A second, more systematic approach, is to encapsulate the mode change actions into components that implement the following interface:

ModeChangeAction
doModeEntryAction(newMode : int) : void   *doModeExitAction(oldMode : int) : void*

Components implementing this interface are called *mode change action components*. A mode change action component is associated to each mode manager and the mode manager calls method doModeExitAction when a mode is exited and method doModeEntryAction when a mode is entered.

Mode change action components take the mode as their argument. In an alternative implementation, a mode change action component is associated to each mode managed by a mode manager and, when it leaves a mode, the mode manager calls doModeExitAction on the associated mode change action component and, when it enters a mode, it calls its doModeEntryAction method.

Mode change actions are a framework hot-spot (the *mode change action hotspot*) since it is through them that applications define the actions to be executed upon mode transitions.

12.3 Coordination of Operational Mode Changes

The operational mode is a property of (some) AOCS components. In order to ensure consistency of behaviour at the system level, it is necessary to ensure that changes in the operational mode of individual components are coordinated. For instance, it is clearly dangerous if the attitude controller is operating in a mode that assumes that high accuracy sensors are powered up and are supplying attitude information while the failure detection component assumes that only basic sensors are switched on. In order to guarantee coordination, the operational mode of a component is exposed as a monitorable property in the sense of section 11.3.2. This gives other components the chance to register their interest in changes in its operational mode and therefore to coordinate their behaviour with it.

12.4 AOCS Mission Mode Manager

Components are responsible for managing their own operational mode. They decide on mode changes based on their observation of the external environment. Part of this information can come from monitoring the state of other AOCS components but part must come from outside the AOCS. The ground station might want to provide information about, for instance, changes in orbital conditions, or the beginning of particular manoeuvres. Similarly, the satellite central computer might send commands to force the AOCS into a survival mode.

The framelet provides a component – the *AOCS Mission Mode Manager* – to supply this information to the rest of the AOCS software. This is a very simple component whose only function is to expose its operational mode to other interested components. Its operational mode is "passive" in the sense that it is set by telecommand and is never changed otherwise. This operational mode plays a special role and is therefore given a special name: *mission mode*. The mission mode is the means through which the ground station or the satellite central computer can signal the need to change operational mode to the AOCS components.

In current AOCS systems, the ground can either force the AOCS to go into a specified mode or it can inhibit certain autonomous mode transitions from taking place. With the proposed, more distributed, mode concept, it no longer makes sense to implement such a direct control over mode management. In principle, it is of course possible to have telecommands that control the mode of individual components but because there may be a large number of mode-dependent components, this may be impractical. It may also be unsafe because changes in the mode of one single component reverberate throughout the AOCS through the property monitoring mechanism.

Essentially, once the decision is taken to give each component responsibility for managing its own mode, direct external control of operational mode becomes impossible. Indirect control is still possible through the mission mode manager but this can only act as a relay of mode-related signals from the ground or the central satellite computer and it is up to the individual components to heed these signals.

As an example consider the common situation where the AOCS software must be endowed with a survival mode that is entered in case some serious anomaly has been detected in an higher-level operational mode. Such a fall-back mechanism can be implemented with the proposed mode concept by endowing the mission mode manager with a survival mode and by arranging for all other mode managers to monitor the mission mode and to treat its transition into survival mode as a symptom that something has gone wrong and that a transition into some kind of safe low-level operational mode is necessary.

12.5 Reusability

This framelet furthers reusability by separating the implementation of the mode-specific algorithms from the mode-switching logic. This makes it possible to construct components that are independent of the operational mode architecture and that therefore have a higher degree of reusability.

None of the constructs offered by the mode management framelet is AOCS-specific. The solution it offers can be reused at both design and architectural level in any other context where the operational mode concept is useful as a modelling and implementation tool. It should however be mentioned that in some domains explicit use of state machines is more common to implement operational-mode dependencies. In such domains, it would be more natural for the framework to directly offer a state machine abstraction.

Table 18 summarizes the constructs exported by the framelet that have been defined or identified in this chapter.

Table 18. Overview of operational mode framelet

Design Patterns	
Mode Management	Separate the mode-switching logic from the implementation of mode-dependent algorithms.
Abstract Interfaces	
`ModeChangeAction`	Interface that characterizes a component encapsulating an action to be taken when a mode manager performs a mode transition.
Mode Manager Interfaces	Interfaces to characterize the mode manager for a specific mode-dependent component.
Core Components	
AOCS Mission Mode Manager	Passive mode manager whose mode is set by tele-command and that serves to relay mode-related signals to other on-board components .
Hot-Spots	
Mode Manager	Applications must define the mode-switching logic for their mode managers.
Mode Implementer	Applications must define the mode-dependent implementations of the strategies associated to each mode manager.
Mode Change Action	Applications must define the actions to be taken when a mode manager performs a mode transition by providing concrete implementations for interface `ModeChangeAction`.

13 The Intercomponent Communication Framelet

An AOCS application instantiated from the AOCS software will be built as a collection of software components. These components interact by exchanging data. In an AOCS context, two major categories of data can be recognized. *Event data* are produced asynchronously by components that wish to signal a change in their internal state or the occurrence of some event. Examples of events include the notification of an error condition, the notification of a unit reconfiguration, or the notification of a successful telecommand execution. *Cyclical Data* (also called *AOCS data*) are instead data that are produced or consumed on a periodic basis by AOCS components. Examples include the output data produced by sensor components, the torque requests produced by the controller component, or the telemetry data produced by the telemeterable components.

The intercomponent communication framelet proposes application-independent mechanisms for handling these two types of exchanges. It offers two design patterns and several core and default components to support their implementation in concrete applications. Both design patterns are based on a shared memory mechanism. Shared memory is used because it has the advantage of decoupling production of data from consumption of data. Its main disadvantage is that data integrity requires access in mutual exclusion. In practice, this problem will seldom arise since AOCS systems – like all embedded control systems – are inherently cyclical and therefore tend to rely on cyclical or otherwise non-preemptive scheduling.

13.1 The Shared Event Design Pattern

This design pattern is introduced to address the problem of allowing components to share access to event reports that are generated asynchronously. The pattern is illustrated in the UML diagram of figure 13.1. Borrowing an idea from the Java language, the events are encapsulated in objects. Both the producer and the consumers of the events have access to a repository component that acts as a shared data area for the exchange of the event objects. The event producer calls method create to ask the repository to create and store a new event object. The event consumer uses the iteration methods to retrieve all the event objects in the repository and process them as necessary.

The instantiation of the design pattern in the framework is supported by two core components AocsEvent and EventRepository that act as base classes

A. Pasetti: Software Frameworks, LNCS 2231, pp. 167-181, 2002.
© Springer-Verlag Berlin Heidelberg 2002

for concrete event and repository classes. Class `AocsEvent` plays the same role in the framework as class `EventObject` does in the JavaBeans event model.

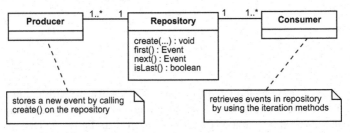

Fig. 13.1. Class diagram of the shared event design pattern

The base class `AocsEvent` defines the three following read-only properties for an event:

- time stamp (the time when the event was created)
- event creator (the component creating the event)
- event identifier (an integer identifying the event type)

Class `AocsEvent` should be subclassed to construct more specific types of events. For example, an event flagging the failure of a failure detection test might include information on the failed test. Since the type and characteristics of events are partly application-specific, the subclassing of class `AocsEvent` defines a framework hot-spot (the *AOCS Event hot-spot*). The framework predefines as default components a number of event classes encapsulating typical events found in AOCS applications. The event subclasses defined by the framework are listed in table 19. Their use will become clear as the description of the framework progresses.

Table 19. Event classes predefined by the AOCS Framework

Event Subclass Name	Event Description
`TelecommandEvent`	Records the execution of a telecommand
`FailureEvent`	Records the detection of a failure
`RecoveryEvent`	Records the execution of a recovery actions
`ManoeuvreEvent`	Records the execution of manoeuvre
`ChangeEvent`	Records change in value of a monitored variable
`ReconfigurationEvent`	Records the execution of a reconfiguration
`ModeEvent`	Records a change in operational mode
`ConfigurationEvent`	Records the detection of a configuration error
`SystemEvent`	Records the occurrence of a system event.

For each type of events, there is an associated event repository. This is preferable to having a single event repository for all events because it helps event consumers to find the events that is of interest to them (normally an event consumer is only interested in events of a certain type). Class `EventRepository` is the base class for all event repositories. It defines the iteration and storage operations in a generic manner independent of the particular type of event. For each category of events of type <EventType>, a class <EventTypeRepository> is offered by the framework (e.g. Events of type `FailureEvent` are stored in instances of class `FailureEventRepository`).

Event producers store events in an event repository and event consumers inspect event repositories to verify whether any events of interest to them have been produced. Event repositories therefore serve two purposes: they act as *factories* of objects of type `AocsEvent`, and they act as *storage points* for objects of type `AocsEvent`. Events by their very nature must be created dynamically. However, in an embedded system, it is undesirable to allocated memory dynamically. Hence, each event repository pre-allocates memory and calls to its `create` method will simply result in a pre-allocated event area being filled with the description of the event. Event repositories expose methods to allow event consumers to iterate through all events currently in the repository. Events are never explicitly destroyed. The repository has a pre-defined capacity and when that capacity is reached, the oldest event is overwritten.

13.2 The Shared Data Design Pattern

This design pattern is introduced to address the problem of allowing components to share access to data that are generated cyclically. The pattern is illustrated in the UML diagram of figure 13.2.

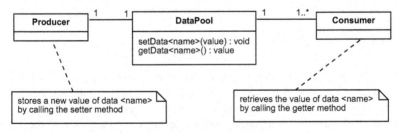

Fig. 13.2. Class diagram of the shared date design pattern

The data pool component contains a single instance of the shared data. The data producer (of which there should be only one: the data owner) calls the setter method to set the data value. The data consumers (of which there can be many) can use the getter methods to retrieve the value of the shared data.

The main problem in instantiating this design pattern is the wide variety of data types that are used in an AOCS. These can be scalars, vectors with 2 or 3 elements representing euler angles or other quantities, quaternions, etc. One solution is to leave the instantiation of the design pattern open. This would then become a design pattern that is *exported* by the framework and that applications must instantiate in a manner that is appropriate to their needs. In designing a framework there is a requirement to reduce variability across applications to the maximum possible extent allowed by the characteristics of the target domain. In pursuit of this objective, the AOCS Framework has tried to introduce a generic data type for all cyclical data. This then makes it possible to instantiate the design pattern within the framework and to associate to it default implementations of data types and data pools.

13.3 AOCS Data

Cyclical data in the AOCS Framework are encapsulated in objects that are derived from the base class AocsData. This is a base type from which the various concrete data types that are used in an AOCS are derived. The subclasses that are provided by the framework prototype as default components are shown in the class diagram of figure 13.3.

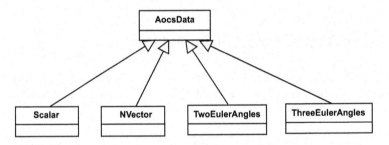

Fig. 13.3. Data types provided as default by the framework prototype

Class NVector represents a generic vector with n components where n is defined at creation time. Class Scalar represents a scalar. Classes ThreeEulerAngles and TwoEulerAngles represents vectors of, respectively, two and three Euler angles. Euler angles are a frequently used measure of the angular displacement of a satellite. Obviously, these data types do not exhaust all possibilities and specific projects can add their own subclasses. The subclassing of class AocsData therefore represents one of the framework hot-spots (the *AOCS data hot-spot*).

Class AocsData is introduced as a generic wrapper for cyclical data. Ideally, it should allow components to process cyclical data in a manner that is independent of their concrete type. Thus, components should be able to perform operations on the abstract AocsData type in the knowledge that this operation would be invisibly dispatched to the correct concrete sub-type. This in particular applies to arithmetic operations: components should, for instance, be able to add together two `AocsData` variables without worrying about whether the addition is implemented as scalar addition, a vector addition, a quaternion addition, etc. This concept was investigated but was found to be impossible to implement without making recourse to dynamic memory allocation. The concept adopted for the AOCS Framework only partially attains the goal of generality of treatment.

Basically, it is recognized that AOCS data present two "faces" to the rest of the AOCS software that reflect the two main purposes for which AOCS components may need to access them. Some components access AOCS data for *housekeeping purposes* like integrity checking, telemetry reporting, failure investigations, etc. Other components access AOCS data for *computational purposes*, namely they use them as inputs or outputs of arithmetic computations. The AOCS Framework allows uniform treatment of data with respect to housekeeping access only. Computational access must instead be done on lower-level data items.

Since housekeeping operations are independent of the specific data type, they are defined as operations directly exposed by class `AocsData` and implicitly inherited by all specific data types. Three types of housekeeping operations are foreseen by the framework: metrics operations, normalization, and time tagging operations.

Metrics operations assume that the concrete data types are defined in a space in which a metric can be defined. The basic metric method is `distance` that computes the distance between two AOCS data. In the case of scalars, this distance is simply the absolute value of the difference between the two items. In the case of vectors and Euler angles, the Euclidean distance is computed. In other cases, type-specific notions of distance must be implemented. The basic method `distance` can be used to build related methods. Examples are:

- Method `equal` returns `true` if the distance is zero.
- Method `size` returns the distance of the object from the zero point in the metric space.
- Method `close` returns `true` if the distance is less than a predefined threshold.

Metrics functions are useful for failure detection as they can be used to perform property monitoring as per chapter 11. Consider for instance the monitoring of an attitude control error. If this error is represented as a variable of `AocsData` type, it will be possible to treat its size as a property and hence to subject it to monitoring to ensure that it remains within certain boundaries.

Normalization operations deserve to be introduced because some data types in common usage on AOCS systems – for instance vectors, quaternions and Euler angles – can be normalized. A method `normalize` is therefore defined that causes the normalization to be performed. In the case of a quaternion or unitary vector, for instance, a call to `normalize` rescales the elements of the data type to ensure that their quadratic sum evaluates to 1. In the case of Euler angles, method `normalize` moves all angles to the interval [-180 deg, +180 deg].

Method `normalize` returns a float indicating how far from the normal range the datum was. This return value can be used to perform consistency checks on data. Thus, for instance, if it was found that the quaternion representing the satellite attitude after integration of its equation of motion has a quadratic sum much larger than 1, then it is likely that the integration algorithm is misbehaving and the fact can be reported as an error.

It is not possible to associate a single time tag to an `AocsData` as this is a composite type that may contain several individual data items (e.g. a `Vector` contains three elements that can be set at different times). Hence, time tagging operations need several methods to access the individual time tags. Useful methods could be `getFirstTimeTag` and `getLastTimeTag` that return, respectively, the oldest and the most recent time tags of the element in the AOCS data composite. Both may be of interest for failure detection purposes since failure to update a certain data may point to an error.

All these methods imply a pure read-only access to a data and do not require any knowledge of its concrete type. They thus realize the objective of generality of treatment of all AOCS data. As mentioned above, this objective cannot be achieved in the case of computational access to the data since computations cannot be done at the abstract level of `AocsData`. The solution proposed by the AOCS Framework is to perform computations upon atomic variables of primitive type. The term *data item* is used to designate one of the elements of primitive type that make up an AOCS data. Unlike AOCS data, data items cannot be further decomposed into lower level entities. A variable of class `ThreeVector`, for instance, contains three data items representing the three elements of the 3-dimensional vector.

Data items are normally private class attributes that cannot be directly accessed from the outside. Access to them must therefore take place through wrapper objects. Two types of wrappers are defined to provide read-only and read-write access, respectively. The two wrappers are represented by classes `DataItemRead` and `DataItemWrite` shown in the UML diagram of figure 13.4.

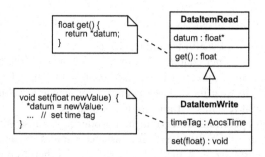

Fig. 13.4. Class structure for data item classes

Attribute `datum` is the reference to the encapsulated data item. A call to method `get` causes the value of the data item to be returned. A call to method `set` (exposed by the `DataItemWrite` class only) can be used to change the data item value. Note that it is assumed that the data item is of basic floating point type and that to each data item a time tag can be associated defining the time when the data item was last updated.

`DataItemRead` objects can be used to establish links between producers and consumers of data. Suppose for instance that component A uses a data item d from component B as an input for its operations. Component B will then expose a method to allow A to get a `DataItemRead` object that encapsulates d. A will then use this `DataItemRead` object to access d. This linking mechanism is illustrated in the implementation case of section 13.7. Class `DataItemRead` is designed to be very light-weight because instances of this class are extensively used in the AOCS Framework.

`DataItemWrite` objects extend `DataItemRead` objects to give write access to the data item they encapsulate. This class adds to the `DataItemRead` class: a reference to the time tag of the data item and methods to set the variable and to read its time tag. The time tag is automatically set by the `set` method.

Bringing together all the above, it is possible to propose the following structure for the base abstract class `AocsData`:

AocsData
metricsOperation(...) : void *normalize() : void* getTimeTag(int) : AocsTime getDataItemRead(int) : DataItemRead getDataItemWrite(int) : DataItemWrite

As already mentioned, this class exposes methods that allow housekeeping operations to be performed directly on `AocsData` variables. Computations are instead done in two stages. First, the data items are retrieved from the `AocsData` variable using one or the other of the two getter methods depending on whether read-only or read-write access is required. Then, the operations are performed on the data items themselves.

All concrete data types – scalar, vectors, quaternions, etc – are derived from the base class `AocsData`. Subclasses will in general only need to override methods `distance` and `normalize` to take account of the class-specific notions of metrics and normalization.

13.4 Data Pools

The data pools are the shared memory areas through which cyclical data are exchanged among framework components. The data pools physically contain the data items. Data users access the data in the pool through references. Only one instance of a data of a certain type can be stored in the pool. This means that if, for instance, the method to set the control torque in the attitude data pool is called twice, then the datum written to the pool at the second call overwrites the datum written at the first call. This mechanism is rigid but safe. It is suitable for cyclical data whose number and type is fixed and can be determined at design time. Note that data pools introduce some overhead because data have to be *copied* to a data pool.

Several classes of data pools may be present where every class groups together logically related data. Multiplicity of data pools makes it easier to subject different classes of data to different treatment. For instance, if it is desired to store attitude data in telemetry, this can be done by simply passing the reference of the attitude data pool to the telemetry manager.

The framework provides a class `AocsDataPool` which defines methods to iterate through the items in the data pool:

AocsDataPool
first() : AocsData next() : AocsData isLast() : bool

Application specific data pools can be derived from this class which gives rise to the *data pool hot-spot*. Derived data pools extend their base class by adding methods to retrieve the individual data in the pool. This is illustrated in the second implementation case example below.

13.5 Implementation Case Example – 1

Consider the implementation case described in table 20. This section shows how it can be worked out during the framework design definition stage. Its objective is to record telecommand executions. This is a typical requirement in an AOCS application where it is usually important that a log be kept of all executed telecommands.

The framework predefines as a default component class `TelecommandEvent` to encapsulate telecommand-related events. The corresponding event repository is also predefined by the framework. Hence, in the spirit of the intercomponent communication framelet, the execution of a telecommand is most naturally recorded as an event of type `TelecommandEvent`.

Telecommands are executed by a telecommand manager component that is one of the core components offered by the framework (see chapter 20). This component needs to be able to create telecommand events. It must therefore have the telecommand event repository as a plug-in component:

```
class TelecommandManager {

    TelecommandEventRepository* tcRep;
    . . .
    // Setter method for TC evt repository plug-in
    void setTcEvtRep(TelecommandEventRepository* rep) {
        tcRep = rep;}
```

```
    // Execution and recording of telecommands
    void activate() {
        . . .                        // execute a telecommand
      if (tcRep!=null)
          tcRep->create(...);        // record execution
    }
}
```

The setter method is used to load the telecommand event repository as a plug-in component. Telecommands are executed within method `activate`. Their execution is recorded by creating a new telecommand event. This is done by calling method `create` on the telecommand event repository. The telecommand attributes are passed as parameters to this method that will then store them in the created event. Components that need to inspect the telecommand log can do so by inspecting the events stored in the telecommand event repository.

Since both the telecommand event repository and the telecommand manager are predefined by the framework, the application developers only need to write the following code to set up the telecommand logging mechanism:

```
// Instantiate the TC event repository
TelecommanEventRepository tcRep;
. . .
// Instantiate and configure the TC manager
TelecommandManager tcManager;
tcManager.setTcEvtRep(&tcRep);
```

This instantiation code will ensure that the telecommand manager records all telecommands executions as events in the telecommand event repository.

Table 20. Implementation case example for event repositories

Objective	Keep a record of all executed telecommands.
Description	An AOCS application must normally log the execution of telecommands. This IC shows how such a log can be constructed.
Framelets	Intercomponent Communication Framelet
	Telecommand Framelet
Constructs	`TelecommandEventRepository` component
	`TelecommandManager` component
Hot-Spots	AOCS event hot-spot
Related ICs	None.
Implementation	- Instantiate a telecommand event repository
	- Plug the event repository in the telecommand manager

13.6 Implementation Case Example – 2

Consider the implementation case described in table 21. This section shows how it can be worked out during the framework design definition stage. Its objective is to allow sensor processing components in an AOCS application to have access to the sensor outputs. This can be done with the shared data design pattern by setting up a data pool that contains all sensor outputs:

```
class SensorPool : public AocsDataPool {
   TwoEulerAngle sunSensorA;
   TwoEulerAngle sunSensorB;
   Scalar gyroA;
   Scalar gyroB;
   . . .
   // Getter methods for data pool entries
   AocsData* getSunSensorA() {
      return &sunSensorA; }

   AocsData* getSunSensorB() {
      return &sunSensorB; }

   AocsData* getGyroA() {
      return &gyroA; }

   AocsData* getGyroB() {
      return &gyroB; }


   . . .
}
```

The example considers a data pool that holds the output of a redundant sun sensor and of a redundant single-axis gyro. The sun sensor output is stored as a vector of two Euler angles and the gyro output is stored as a scalar. The getter methods that give access to data pool entries return objects of generic type AocsData. This is essential to ensure that all data pool entries are treated uniformly by their users. The getter methods only return references to the data pool entries. Through these references the data pool entries can be either read or written to. The mechanism for doing this is illustrated in the next implementation case.

Table 21. Implementation case example for data pools

Objective	Give sensor processing components access to sensor outputs.
Description	The processing of sensor outputs is one of the primary functions in an AOCS application. This IC shows how sensor outputs can be made potentially available to all AOCS components.
Framelets	Intercomponent Communication Framelet
Constructs	AocsDataPool component
	AocsData abstract class
Hot-Spots	Data pool hot-spot
Related ICs	This IC is extended by the IC in the next section.
Implementation	Subclass AocsDataPool to create a data pool containing all sensor outputs.

13.7 Implementation Case Example – 3

Consider the implementation case described in table 22. This section shows how it can be worked out during the framework design definition stage. Its objective is to allow a component to process the outputs of a sun sensor. Processing of sensor outputs is a common task in an AOCS application.

The sun sensor output is assumed to be stored in a data pool as illustrated in the previous implementation case. The component processing the sensor output is defined as a component that can process two generic inputs:

```
class Component {

    DataItemRead inp1, inp2;
    . . .

    // Set up the link to the data source for the 1st input
    void linkInput1(DataItemRead d) {
        inp1 = d;
    }

    // Set up the link to the data source for the 2nd input
    void linkInput2(DataItemRead d) {
        inp2 = d;
    }

    void activate() {

        float u1 = inp1.get();
        float u2 = inp2.get();
        result = f(u1,u2);              // process inputs
        . . .
    }

}
```

This component processes two inputs that it sees as read-only data items. If it is to be made to process the sun sensor outputs, then it should be instantiated and configured as shown in the following pseudo-code:

```
// Sensor data pool reference (see previous IC)
SensorPool* sensorPool;

// Retrieve the sun sensor output from the data pool
AocsData* ss = sensorPool->getSunSensorA();
. . .

// Instantiate and configure the data processing component
Component* component = new Component();
component->linkInput1(ss->getDataItemRead(1));
component->linkInput2(ss->getDataItemRead(2));
```

The last two statements create a permanent link between the data processing component and the sun sensor outputs: the data processing component is handed the data items that encapsulate the references to the data it must process. The use of the data item mechanism allows the source of the data to be separated from its processing. The same data processing component could be used without changes to process any other data pool variables. The framework uses this data linking mechanism wherever components have to process data that are generated outside themselves.

Finally, the pseudo-code example does not consider the issue of concurrency. It implicitly assumes a non-preemptive scheduling policy. If this were not the case, then the copying of the sensor data should be done in a protected section:

```
void activate() {

    /* start non-preemptable section   */
    float u1 = inp1.get();
    float u2 = inp2.get();
    /* end non-preemptable section    */

    result = f(u1,u2);              // process inputs
    . . .
}
```

This code ensures that the two input data are consistent in the sense of being acquired from a data pool that cannot change its internal state in between the two acquisitions.

Table 22. Implementation case example for data item

Objective	Let a component process the output of a sun sensor.
Description	One function of an AOCS application is to process sensor outputs. This IC shows how a sensor processing component can be linked to the sensor outputs.
Framelet	Intercomponent Component Framelet
Constructs	AocsData abstract class
Hot-Spots	None
Related ICs	This IC extends the IC of the previous section.
Implementation	- Retrieve the sun sensor output from its data pool
	- Retrieve the data items from the AocsData representing the sensor output
	- Load the data items into the data processing component

13.8 Alternative Implementations

The data exchange design patterns proposed by this framelet are based on the use of shared memory. An alternative design solution to the problem of transferring data between components could be a message passing mechanism where the information transfer is mediated by a transmission channel over which a data packet is transferred that contains the information item. This mechanism is the most general and has been used for instance in a frameworks for avionics system described in [33] but it also carries the heaviest overhead. It would only be useful if the AOCS software were distributed which is not the case at present.

Much of the interaction between AOCS components consists of command requests that are sent by a component to another to force a change in the latter's internal state. At present, command requests from a component A to a component B are implemented in the framework as calls performed by A upon methods exposed by B. Thus, for instance, to the command to reset a component there must correspond a `reset` method that must be called on the component to be reset: if component A wishes to send a "reset" command to component B, it will call its `reset` method. In an alternative implementation, command requests could be packetized as events to be transferred using the normal event transfer mechanism. This approach would have the advantage of confining the execution threads of active components within their own boundaries but would introduce a delay between the time when the command is issued (i.e. the time when the command event is deposited in the shared data area) and the time when the command is executed (i.e. the time when the command is picked up from the shared memory area). It would also cause both memory and execution overheads.

Shared data repositories could in principle be either *active* or *passive*. In the case of active repositories, clients register their interest in a certain item and are notified when the item is updated. The proposed design patterns prescribe a passive concept. This is uncontroversial in the case of the cyclical data that are generated at known times and whose consumers need to access them periodically. A notification mechanism seems instead more natural for event data. In the AOCS Framework, this was avoided in order to have maximum decoupling between event producers and event consumers. Additionally, since the frequency with which events are expected to be generated in an AOCS application is rather low, the overhead arising from the need for event consumers to inspect the event repositories to check whether there is anything of interest for them is very low. The division of events into categories assigned to different event repositories also helps keep this overhead low because event users can minimize the inspection overheads by concentrating only on the event repositories of interest to them.

The use of data item objects to perform computations on the cyclical data might appear overcomplex. In a simpler alternative design, one could endow `AocsData` with the following methods:

```
void set(int j, Real newValue);
Real get(int j);
```

These methods could be used to directly set and get the j-th component of the datum. Use of a wrapper is preferred because it makes it easier to associate attributes, like the time tag, to the data items. At present, there is no access control to data items. In an extension of the framework, data items could be given an owner which is the only component that is allowed to have write access to them. This could be a second attribute of data items and further reason for encapsulating them in dedicated objects.

Data items also simplify the linking of a component to its data. If, for instance, a component needs to update a given datum on a regular basis, it can be given as part of its initial configuration, a copy of the data item object that encapsulates the reference to the datum itself. This creates a permanent link between the component and the data upon which it operates. In the AOCS Framework, this type of linking mechanism is notably exploited in the implementation of the control channel components introduced in the next chapter.

The property objects introduce in section 11.1 encapsulate variables and provide read access to them. They are in this respect similar to data items and, at first sight, one could think of using data items in lieu of property objects (or vice versa). This was avoided because the intended usage for the two types of objects is very different. Data items are meant to facilitate linking components to their data sources and will therefore be used extensively in any AOCS application. They must therefore be light-weight and efficient. By contrast, property objects are meant for monitoring purposes and may have to contain other information in addition to the bare value of the property (e.g. a property identifiers, an identifier of the property type, etc).

13.9 Reusability

This framelet enhances reusability in three ways. Firstly, through the use of shareable data areas for inter-component communications, it decouples the *production* of data and events from their *consumption*. Secondly, by allowing uniform treatment of all data types, it makes component interfaces independent of the type of data they process. Thirdly, with the *data item* concept, it provides a way to link components together at run-time. Flexibility in reuse is guaranteed by the possibility of creating new subclasses of `AocsData` and `AocsEvent` to encapsulate new types of data or events.

The distinction between event and data is common to embedded control systems in general and hence this framelet could be ported to other embedded control systems. However reuse would probably be at the design level only. The specific architecture proposed here – the structure of the `AocsEvent` and `AocsData` classes – are tied to the AOCS and may need modifications before being applicable in other domains.

Table 23 summarizes the constructs exported by the framelet that have been defined or identified in this chapter.

Table 23. Overview of intercomponent communication framelet

Design Patterns	
Shared Data Pattern	Allow components to have access to events generated asynchronously.
Shared Event Pattern	Allow components to have access to cyclical data.
Abstract Classes	
AocsData	Define an interface to access a generic data type.
Core Components	
AocsDataPool	Base class for all data pools.
AocsEvent	Base class for all categories of events.
EventRepository	Base class for all event repositories.
DataItemRead	Encapsulation of a read-access to a data item
DataItemWrite	Encapsulation of a write-access to a data item
Default Components	
Event classes	Encapsulation of commonly used event categories: telecommand events, failure events, reconfiguration events, system events, etc.
Concrete data classes	Encapsulation of commonly used data types: scalar, n-vector, 2-Euler angles, 3-Euler angles, etc
Hot-Spots	
AOCS Event	Applications must define their event types by providing their own subclasses of AocsEvent.
AOCS Data	Applications must define their data types by providing their own subclasses of AocsEvent.
Data Pool	Applications must define their data pools by providing their own subclasses of AocsDataPool.

14 The Sequential Data Processing Framelet

An AOCS software must normally perform a significant amount of data processing. In the AOCS Framework data that are passed through several consecutive stages of processing are said to go through a *sequential data processing chain.* As an example of a sequential data processing chain consider the processing of the raw data collected by a sensor:

- conversion from raw data to engineering units;
- correction for sensor misalignments;
- bias correction;
- transformation from sensor to satellite reference frame

In another example, an attitude controller processes the attitude measurement as follows:

- filtering of attitude data to remove the measurement noise;
- filtering to cut off frequencies that might excite solar panel oscillations;
- passage through a controller to compute the control torque.

Each bullet can be seen as representing a processing stage in a sequential data processing chain. In these examples, each processing stage is single-input-single-output but a processing chain can include multi-input-multi-output stages as in the case shown in figure 14.1. The figure shows a 2-axis controller where the sensor measurements are first passed through a Kalman Filter and are then fed to an attitude controller.

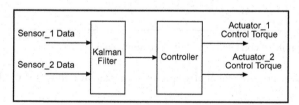

Fig. 14.1. Example of a sequential processing chain

In conventional AOCS applications, no standardized mechanism is defined to encapsulate data processing. Data processing algorithms are simply hard-coded in dedicated procedures inside the objects that need them. Since these algorithms are clearly a hot-spot in the AOCS domain, the framework should devise some way of

A. Pasetti: Software Frameworks, LNCS 2231, pp. 183-192, 2002.
© Springer-Verlag Berlin Heidelberg 2002

handling generic data processing algorithms. This is the design problem that is addressed by the sequential data processing framelet.

14.1 Control Channels

In considering the implementation of sequential processing chains, a trade-off must be made between generality and complexity: the most general implementation is also the most complex and the simplest implementation cannot cover all cases of processing chains. In view of the wide variety of data processing chains that can conceivably be used by AOCS applications, the framework has opted for generality of treatment. It introduces the *Control Channel* concept that covers the most general case of multi-input-multi-output processing stages linked in any arbitrary manner.

A sequential processing chain is made up of inter-connected processing blocks. The term *control channel block* (or simply *block*) is used to designate a processing block that cannot be further decomposed into lower level processing blocks. Blocks can be concatenated in chains as shown in figure 14.2 where arrows represent data flow: the user input is fed to block 1, and the user output is taken from block 4. Block chains can be nested within higher-level blocks called *control channel superblocks* or simply *superblocks*. Superblocks and blocks can be freely mixed in processing chains as in the example in figure 14.3. The entire control chain is enclosed in a superblock (superblock 3) with one input and two outputs. The superblock contains two superblocks (superblocks 1 and 2) and one simple block (block 6).

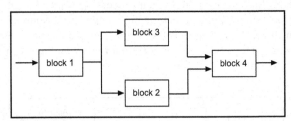

Fig. 14.2. Example of concatenated control channel blocks

In general, a control channel implements a transfer function of the following kind:

$$x_{t+\Delta t} = f(x_t, u_t)$$
$$y_t = g(x_t, u_t)$$

where the usual notation is adopted with u representing the input vector, y the output vector and x the state vector. The fundamental operation to be performed on a control channel is the propagation of its output signal from time t to time (t+Δt). A *propagate(t)* operation is defined on control channels that causes their outputs to be propagated up to time t. The time t to which the output values are propagated is called the *validity time* of the outputs.

In order to compute y(t), the control channel needs to know u(t). Thus, if control channels are arranged in a sequential chain, a propagation request must be passed to upstream blocks. For example, consider again the control channel chain of figure 14.2. Suppose `propagate(t)` is called on block 4. Before this operation can be performed on this block, it is necessary to update the inputs to the block and this is done by calling the same operation `propagate(t)` on blocks 2 and 3. Thus, propagate requests percolate backward along the processing chain. Eventually, they will reach a control channel that takes its inputs from a component that is not itself a control channel. Such external signals will be assumed to be fed to a control channel using a *zero-order hold*. This means that their value will be assumed to be constant across propagation instants.

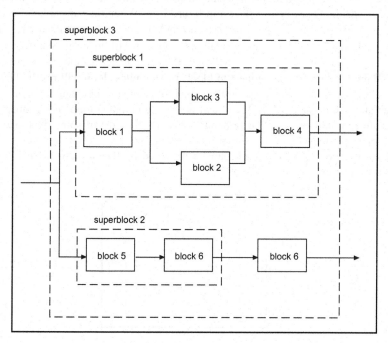

Fig. 14.3. Example of control channel superblock

To illustrate the zero-order hold concept, consider for instance the situation in figure 14.4. Suppose that both input and output have the same validity time t. Suppose now that operation `propagate(t+Δt)` is called on the control channel. The control channel will respond by updating its output to y(t+Δt) and it will do so by assuming that the input remains constant and equal to u(t) throughout the propagation interval. If a first-order hold were instead used, then the value of the input signal would be computed by linear extrapolation from its last two known values. Higher order holding systems are also possible. Zero-order holding is, however, by far the most commonly used type of holding mechanism in satellite control systems and is the only one for which the AOCS Framework makes provisions.

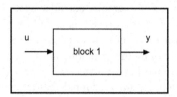

Fig. 14.4. Example of control channel block

Note that the propagation mechanism, as it stands, cannot cope with signal loops. Consider the situation shown in figure 14.5. When the super block receives a `propagate` request, it will route it to block 1. Before executing the `propagate` action, block 1 will try to update its inputs. It will do so by issuing a `propagate` request to block 2. This will in turn try to update *its* inputs and will do so by issuing a `propagate` request to block 1. An endless loop will result. This example can be generalized to show that signal loops in control channel connections, either at block or super block level, will result in an endless loop. The propagation mechanism could be modified to handle loops (at the cost of some overhead) but this is judged unnecessary as signal loops should not arise in an on-board control system. The control channel mechanism provided by the framework will simply fail in the presence of signal loops.

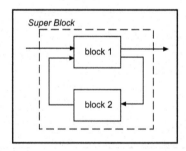

Fig. 14.5. Control channel blocks connection with signal loop

14.2 The Control Channel Design Pattern

This design pattern is introduced to model the control channel concept and in particular the distinction between blocks and super blocks. The control channel design pattern is obtained by refining the composite pattern of [44] and it is shown in the UML diagram of figure 14.6.

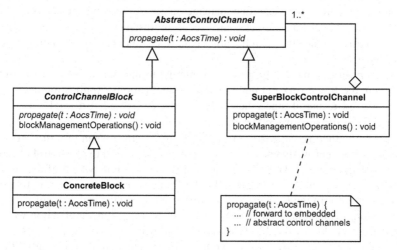

Fig. 14.6. Class diagram of control channel design pattern

AbstractControlChannel is a pure interface that exposes the methods that are common to all control channels, regardless of whether they are blocks or superblocks. Control channels are always seen by their clients as instances of this interface. Since both blocks and superblocks are derived from AbstractControlChannel, they can be treated in a uniform manner as instances of abstract control channels. The fundamental operation exposed by AbstractControlChannel is propagate(t) that propagates the input signals to the outputs up to time t as discussed in the previous section. This method gives rise to the *control channel hot-spot* through which application developers must define the transfer function to be implemented by the control channels they use.

Abstract control channels can be implemented either as control blocks or as control super blocks. ControlChannelBlock is an abstract class that acts as base class for all concrete control blocks. It provides concrete implementations of methods and data structure to manage block operations. This class for instance provides data structures to buffer the input and output signals and operations to reset them. Such data structures and operations are common to all control channel blocks. This class is abstract because it does not provide any implementation for method propagate. The implementation of this method defines the concrete transfer function that is implemented by the control channel. Concrete control blocks specialize ControlChannelBlock by providing concrete algorithms for the propagation of the input signals through the control block. The AOCS Framework provides as default components concrete implementations of this class that implement common transfer functions such as PD blocks, PID controllers, integrators, etc.

Super blocks are represented by instances of class `ControlChannelSu-perBlock` which is derived from `AbstractControlChannel`. This class defines a default component that manages a set of interconnected lower-level blocks. The lower-level blocks are seen as instances of `AbstractControlChannel` since they can be either control channel blocks or control channel super blocks.

The control channel design pattern is internal to the framework: it is instantiated within the framework but not exported to applications. Application developers will not need to use it since they only see the `ControlChannelBlock` and `ControlChannelSuperBlock` classes.

Calls to method `propagate()` can be recursive since when they are called on a given component A, they have to be propagated backward to all control channels directly or indirectly linked to A's inputs. This recursion is typical of the composite design pattern and it is an example of a framework design pattern introducing a construct with timing properties that, from a purely syntactical point of view, are statically unpredictable. This presents problems in a real-time application where it is usually desirable to be able to analyze schedulability statically. This is an example of the problem discussed in section 9.10. the only solution is to provide semantic information to bound the depth of recursion. In the case at hand, this is easy to do since the maximum depth of the recursion is given by the maximum length of a chain of connected control blocks.

Control channels transform input values into outputs values. In the AOCS Framework, they return their outputs as data items, namely they return objects of type `DataItemRead` that encapsulate a reference to the output itself. Their inputs can be linked either to external variables or to the outputs of other control channels. Input linking is also done through data items. Thus, when the input of a control channel needs to be linked to a particular external variable, it is passed the `DataItemRead` object that encapsulates the desired variable. If instead it needs to be linked to the output of another control channel, it is passed the data item returned by the source control channel to encapsulate its output. This is an example of how concrete constructs from different framelets – in this case the sequential data processing and the component communication framelets – are merged together at implementation level.

The terms "block" and "superblock" that designate, respectively, simple and composite control channels, are borrowed from the Xmath world. Blocks and superblocks as defined here map to the homonymous concepts in Xmath. This is not just a coincidence. Xmath can be used to design and simulate the transfer functions that are commonly used in control systems and that are then implemented in the sequential data processing chains. An important part of its power and appeal is its ability to automatically generate subroutines that implement the transfer function defined by the user. As was argued in section 3.3, there is a relationship of complementarity between tools like Xmath and the AOCS Framework. At architectural level, this complementarity appears as a hot-spot where code generated from an autocoding facility can be embedded within the AOCS Framework. In order to demonstrate this complementarity, the AOCS Framework offers a default component that, from the framework's point of view, is a control channel (it implements interface `AbstractControlChannel`) but that also acts as a wrapper for a routine generated by the autocode tool of Xmath. This then allows embedding of Xmath models into applications instantiated from the framework and their mixing with control channels that have been obtained in other manners, perhaps by manual coding or perhaps using other autocoding facilities. Similar wrappers could also be developed for Matlab although this was not made in the AOCS Framework project.

14.3 Implementation Case Example

Consider the implementation case described in table 24. This section shows how it could be worked out during the framework design definition stage. Its objective is to construct a component to implement a proportional-derivative, or PD, control action. PD controllers are widely used in embedded control systems in general and in AOCS applications in particular.

A PD controller – or indeed any kind of controller – can be seen as a sequential data processing block and, in terms of the abstractions defined by this framelet, it is best implemented as a control channel block. The framelet prescribes that concrete control channel blocks be implemented as instances of classes derived from the abstract class `ControlChannelBlock`. This class is provided by the framework as a base abstract class. Its definition could be as follows:

```
class ControlChannelBlock {

    float[] input;
    float[] output;
    float[] state;
    . . .

    void propagate(float t)=0;

    . . .    // block management operations
}
```

The `input`, `output` and `state` arrays represent arrays that are predefined by the base class to hold the latest set of input and output values and any internal state variables. Their size is defined at run-time when the control channel component is configured. The pure virtual method represents the hot-spot where application developers can insert the application-specific propagation algorithm. The `ControlChannelBlock` class additionally defines and implements methods to handle the loading of the inputs, their linking to external data sources, the management of the outputs, etc. All these functions are independent of the propagation algorithm and can therefore be defined at the level of the base abstract class for all control channels.

A concrete control channel implementing a PD transfer function can be implemented as follows:

```
class PD_Block : public ControlChannelBlock {

    float kp;      // proportional gain
    float kd;      // derivative gain
    float old_t;   // last activation time

    void propagate(float t) {
        output[0]=input[0]*kp+
                     kd*(input[0]-state[0])/(t-old_t);
        state[0]=input[0];
        old_t=t;
    }
}
```

The state variable is used to hold the previous activation value. The derivative term is computed in a very naïve way as the ratio between the increment in the input and the increment in the time. The important point to note is that the definition of a concrete control block can be done by overriding a single method. All the other methods are inherited from the base control channel class. This makes it very easy to construct components implementing application specific transfer functions and gives all such components uniform external interface because they all implement the same base class.

Components, like the PD controller described here, that implement common types of transfer functions should be provided by the framework as default components.

Table 24. Implementation case example for control channel

Objective	Construct a PD controller.
Description	A PD controller is a common type of controllers used in AOCS applications. This IC shows how a PD controller can be built as a control channel component.
Framelet	Sequential Data Processing Framelet
Constructs	ControlChannelBlock abstract class
Hot-Spots	Control Channel
Related ICs	None
Implementation	Subclass ControlChannelBlock to override its propagate method to implement the PD transfer function

14.4 Alternative Solutions

Control channels can be used to implement a generic combination of blocks and superblocks. If only linear (as opposed to tree-like) processing chains were considered, then the application of successive processing stages to a set of data could be seen as a form of *decoration* in the sense of the decorator pattern of [44]. This restricted type of control channels could therefore be implemented using the decorator pattern. Both the decorator and the composite solutions to the sequential data processing problem model a data flow system as made up of interconnected and nested blocks. All blocks are treated as homogeneous since they are derived from the same base class (`AbstractControlChannel` in the case of control channels and a base `Component` class if a decorator solution were adopted). There is therefore no syntactic restriction to how blocks can be linked to each other or nested within each other. An interesting alternative is to view a particular data flow system as a *sentence in a simple formal language*. In the AOCS context, the basic elements of the language could be blocks like:

- real AOCS sensor
- fictitious AOCS sensor[18]
- real AOCS actuator
- fictitious AOCS actuator
- controller
- filter
- state estimator

These blocks are the 'words' of the 'sentence'. They can be strung together according to a small set of formal rules. Examples of rules (expressed in informal language) could be:

- a sensor must be followed either by a fictitious sensor or by: a filter, or a state estimator, or a controller;
- a controller can only be followed by: a filter, or an actuator, or a fictitious actuator;
- a fictitious actuator can only be followed by another fictitious actuator or by a real actuator.

These and other similar rules could be easily formalized to create a grammar specifying how data flow blocks can be linked together. In the present context, a 'sentence' would be a data flow diagram for the AOCS and the process of 'evaluation' would refer to the processing of data that are input at one end of the data flow diagram (the 'sensor end') and result in outputs being produced at the other end of the data flow diagram (the 'actuator end').

This formal language approach is similar to that proposed in [10, 11] for the control data flow part of avionics systems. It is very attractive for its generality and ease of extension. It would result in components being made available to designers that can only be linked together in manners that are semantically meaningful. From an implementation point of view, however, this approach would be significantly more complex than the approach baselined in the AOCS Framework because it would require the development of compiler-like tools to check the validity of component combinations proposed by the user.

---

[18] The concept of fictitious AOCS units will be covered in section 15.5

14.5 Reusability

The architecture proposed here for control channels furthers software reusability because it provides a standard interface for components that must perform sequential data processing. The processing algorithms are embedded in the processing components and the latter are handled through the standard interface in a manner that is independent of the particular algorithm they contain. Additionally, the option is provided of linking individual blocks and nesting them within higher-level blocks thus allowing complex data processing chains to be formed by combining simple basic blocks.

The use of sequential processing chains is very common in embedded control systems and this framelet could be reused in embedded control domains other than the AOCS. Reuse could be down to code level because the control channel block base class and the control channel super block component are domain-independent.

Table 25 summarizes the constructs exported by the framelet that have been defined or identified in this chapter.

Table 25. Overview of sequential data processing framelet

Design Patterns	
Control Channel	Allow all control channels to be treated uniformly regardless of whether they represent individual control blocks or chains of interconnected and nested control blocks.
Abstract Classes and Interfaces	
AbstractControlChannel	Interface for a generic control channel.
ControlChannelBlock	Base class for control channel blocks that defines all block operations but leaves the implementation of the block transfer function open.
Core Components	
ControlChannelSuperBlock	Control channel that can contain any combinations of other control channels.
Default Components	
Concrete Blocks	Implementation of commonly used transfer function blocks: PD controller, PID controller, integrator block, etc.
Xmath Wrapper	Wrapper component for a routine generated by the autocode tool of Xmath to represent a generic transfer function modelled within the Xmath environment.
Hot-Spots	
Control Channel	Applications must define the transfer functions implemented by their control channels. At control block level, this is done by providing implementations for method propagate in class ControlChannelBlock. At super block level, this is done by chaining together control blocks.

15 The AOCS Unit Framelet

One of the distinctive features of embedded systems is that they need to control external hardware. In the case of embedded control systems, this consists of external sensors and actuators. The AOCS Framework uses the term *AOCS units* to designate a piece of equipment that is external to the AOCS computer but is part of the AOCS system.

Conventional AOCS software standardizes management of external units – if at all – only at the very lowest level by using standard routines to handle the hardware interfaces to the external equipment. This framelet tries to go further. The main problem is the wide variety of external hardware. This dictates an approach already adopted in other frameworks for embedded systems [63] that is based on the separation between unit management and other AOCS functionalities. Separation is achieved by having the units represented within the AOCS software by proxy components that encapsulate the interactions with the hardware and that implement a generic unit interface. The interface acts as an adapter (in the sense of the adapter design pattern of reference [44]) between the AOCS software and each AOCS unit.

Use of proxy components when dealing with external entities is a classical solution most notably adopted in the DCOM and CORBA architectures. Characterization of external equipment through a standardized interface is also a known solution for embedded control systems. It was for instance used in the ADAGE project [10, 11]. ADAGE however has several interfaces for different classes of units whereas the AOCS Framework uses the same interface for all sensors and actuators. Restriction to a single common interface is an advantage because it simplifies the management of the units but it is also a weakness because it restricts the range of units that are covered by the framework.

One innovation of the AOCS Framework is to split this common unit interface into two interfaces to recognize that, in a control system, there are two distinct modes of access to external units. This recognition is a result of the wider scope of the AOCS Framework that encompasses not just the flow of control data from sensors trough the controller to the actuator but also covers housekeeping functions like failure detection, failure recovery, reconfiguration management, data logging, etc.

AOCS units can potentially be very different ranging from very simple sun presence detector to very complex devices like GPS receivers or autonomous star trackers. In accordance with the principle laid down in section 9.1, the concept proposed here for unit management is targeted at units of the kind commonly in use at present. Active units (see section 2.1.3) are thus excluded. Their inclusion

A. Pasetti: Software Frameworks, LNCS 2231, pp. 193-208, 2002.
© Springer-Verlag Berlin Heidelberg 2002

would require a substantial addition to the framework probably requiring either the development of unit-specific interfaces or the development of abstractions to encapsulate data exchanges with the active units along the lines of the abstractions proposed in chapter 20 to model telecommands.

The AOCS unit framelet is built around a major hot-spot representing the specific characteristics of the external unit. It associates two abstract interfaces to this hot-spot and additionally offers a major design pattern to handle components with unit-like behaviour (the so-called "fictitious units"). Mainly for testing purposes, it also offers default components representing proxies for simple units.

15.1 Abstract Unit Model

The design problem addressed by the AOCS unit framelet is to devise a way of handling external units in a uniform manner. This problem can only be solved if some constraints are imposed upon the variability of the units to be thus handled. The abstract unit model describes the generic unit that underlies the handling model proposed by the framelet. Its first assumption is that units are *passive*, namely incapable of initiating data exchanges with the AOCS computer or with other units. All data transactions with a unit occur in response to a command from the AOCS computer. Data transactions between AOCS computer and AOCS units fall under two categories:

- *Functional Transactions*: This type of transaction is cyclical and relates to the primary function for which a unit was designed. Thus, for instance, a thruster unit periodically receives firing profiles, a gyro periodically supplies rate information, a reaction wheel periodically receives torque requests.
- *Housekeeping Transactions*: This type of transaction may be occasional and it consists of commands sent by the AOCS computer to the unit (e.g. to change its operational mode, to switch the unit on or off, etc) and of housekeeping data sent by the unit to the AOCS computer (e.g. temperature readings, results of self-tests)

The status of a unit vis-à-vis the AOCS software is defined by the following four attributes:

- *Power status:* Units can either be "power on" or "powered off"
- *Operational Mode*: Units may be in one of several operational modes
- *Health Status*: Units can either be "healthy" or "unhealthy"
- *Self-Test Result*: Units may be able to perform a *self-test*. The result of a self-test is an integer code.

AOCS units are represented within the AOCS software by proxies that implement a standard interface. The proxy components are called *AOCS Unit Proxies*. Data within these components can exist at two levels of abstraction:

- the *raw data level* representing the data as they are transmitted on the physical communication link between the unit and the AOCS computer;
- the *AOCS data level* representing the data as they are used by clients of the AOCS unit proxies in the AOCS software. Such data are expressed in engineering units and, if they depend on reference frame, are expressed in the spacecraft reference frame.

The raw data level is mission-specific and cannot be defined here. This data level, however, is internal to the AOCS unit proxies. All exchanges between the AOCS unit proxies and other AOCS software components take place at the AOCS data level. It is the responsibility of the AOCS unit proxy to perform conversions between raw and AOCS data levels.

The model for the exchange of data between hardware and AOCS unit proxies is illustrated in figure 15.1. The figure assumes the case of a unit that can both receive and send data (this could for instance be the case of a reaction wheel that receives torque requests and sends wheel velocity measurements). Many units will only work in receive or send mode. As shown in the figure, the AOCS unit proxy maintains *hardware buffers* where incoming or outgoing data are deposited at raw data level. The transfer between these buffers and the hardware interface is done by low level mechanisms.

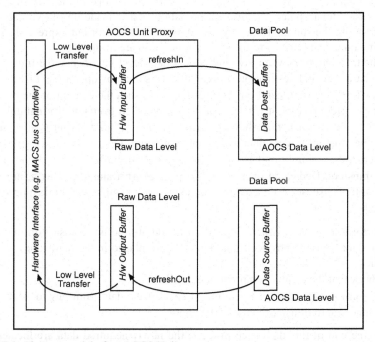

Fig. 15.1. Abstract model of an AOCS Unit

The AOCS on satellites developed by the European Space Agency is often built around a dedicated bus called the MACS bus. In a typical example, therefore, the hardware interface could be a MACS bus controller. Data reception triggers an interrupt whose servicing routine deposits the incoming data word in the hardware input buffer where it remains available for further processing by the AOCS unit. Data to be sent to the unit must instead be written to an I/O address. In that case, the hardware output buffer serves as the source for the data written to the I/O port by a dedicated routine.

Only single hardware buffers are shown in the figure. In reality, the buffers may be made up of several memory locations holding related data. The hardware output buffer for a 2-axis sun sensor, for instance, will consist of at least two locations holding the two sun angle measurements.

The data that the units send out must ultimately come from other components in the AOCS software. Similarly, the final destination of the data that a unit receives is some data consumer component within the AOCS software. According to the concept for data exchanges among AOCS components presented in chapter 13, unit proxies take their inputs and deposit their outputs in data pools. An AOCS unit proxy therefore maintains links to data pool locations where the incoming and outgoing data are stored as variables at AOCS data level. The location where incoming data are stored is called *destination data buffer* and the location from which outgoing data are retrieved is called *source data buffer*. The destination and source data buffers are sets of data items in the data pools. In the case of a 2-axis sun sensor, for instance, the destination data buffer is made up of two data items representing the sun angles as measured by the sun sensor but expressed in engineering units and referred to the spacecraft coordinate frame.

The translation between raw and AOCS data level, and hence the transfer between hardware and destination/source buffers, is done by operations refreshIn and refreshOut. Operation refreshIn refreshes the content of the destination data buffer updating it with the latest data received from the external unit. Operation refreshOut refreshes the hardware output buffer ensuring that it contains the most up-to-date data to be sent to the unit. Operations refreshIn and refreshOut simply update the content of certain buffers. They do not initiate data transfers. Dedicated operations, acquireData and sendData, are provided for this purpose. A data acquisition cycle, for example, would then proceed as follows:

- Perform a acquireData operation to initiate the bus transaction to acquire the data from the physical unit. The data are placed in the hardware input buffer.
- Check that the acquisition process is finished
- Perform a refreshIn operation to convert the data from raw to AOCS data format and to transfer them to a data pool.

At the end of this three-step process, the newly acquired data are located in a data pool in AOCS data format and referred to the satellite reference frame. From the data pool, they are accessible to all other components in the AOCS software.

15.2 The `AocsUnit` Class

AOCS unit management is based on the introduction of an abstract base class –
`AocsUnit` – that defines a generic interface for exchanges with concrete units
and encapsulates an AOCS unit proxy conforming to the model described in the
previous section. This class implements two interfaces: `AocsUnitFunctional`
and `AocsUnitHousekeeping`. The two interfaces represent two conceptually
different types of access to the unit. Interface `AocsUnitHousekeeping` represents the commanding interface of the AOCS unit. Through it, it is possible to perform operations like resetting the unit, checking its health status, getting its bus
address and performing other housekeeping services.

An AOCS unit is seen by the rest of the AOCS software as a server offering
certain services (e.g. a gyro is a component that can be queried for rate information). Interface `AocsUnitFunctional` models this high-level functional interface.

The two interface `AocsUnitFunctional` and `AocsUnitHousekee-
ping` could in principle be merged together. They are kept separate to allow the
definition of *fictitious units*, namely objects that, though not true AOCS units,
mimic their functional behaviour (see below). Obviously, both `AocsUni-
tHousekeeping` and `AocsUnitFunctional` only capture behaviour that is
common to all AOCS units.

Concrete AOCS unit classes are derived from `AocsUnit` through class inheritance. The overall class diagram for AOCS units is shown in figure 15.2

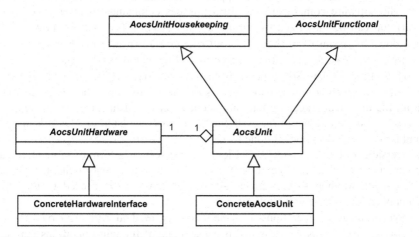

Fig. 15.2. Class diagram for the AOCS unit concept

Concrete unit proxies delegate low level operations to instances of class
`AocsUnitHardware`. Class `AocsUnitHardware` encapsulates the direct exchanges with the hardware. Its instances are normally not visible to the clients of

the units within the AOCS software. Several AOCS units may share access to the same AOCS unit hardware instance.

The definition of unit-specific functionalities by subclassing AocsUnit represents a framework hot-spot (the *AOCS unit hot-spot*) as does the definition of the hardware interface with the units by implementing interface AocsUnitHardware (the *AOCS hardware unit hot-spot*).

15.2.1 The AOCS Unit Housekeeping and Functional Interfaces

The AocsUnitHousekeeping interface is defined, at the design level, as follows:

AocsUnitHousekeeping
initialize() : void
selfTest() : void
resetUnit() : void
resetTransaction() : void
acquireHousekeepingData() : void
refreshHousekeepingIn() : void
ioLinkOperations() : void
synchronizationOperations() : void
powerControlOperations() : void
setHousekeepingInConverter() : void

Units can be reset by calling method resetUnit (note that this is different from the generic reset method that applies to all AOCS components: the latter resets the software component, the former issues commands to reset an external unit) and can be initialized by calling initialize. They can be powered up and down by means of methods powerControlOperations. Some units can perform self-tests. A self-test is initiated by calling method selfTest.

Housekeeping data are acquired by calling acquireHousekeepingData. A call to this method initiates the process for the acquisition of housekeeping data from the unit. Housekeeping data are acquired at raw data level and placed in a hardware buffer internal to the AOCS unit. Their conversion to AOCS data level and transfer to a data pool is done by calling refreshHousekeepingIn. The conversion from raw to AOCS data level is best seen as a form of sequential data processing (see chapter 14) and can then be delegated to a control channel plug-in component. Method setHousekeepingInConverter is used to load this component during the unit configuration.

Synchronization operations – symbolized by method synchronizationOperation – must be provided because some of the unit operations may extend over time and it may be desirable to implement them as non-blocking method calls. For instance, performing a self-test may require some time. Method selfTest is then best defined as simply starting the self test (non-blocking method call). A separate synchronization method is required to check whether the self-test has terminated and still another method is needed to retrieve the results of the self-test.

Linking operations – symbolized by method `ioLinkOperation` – are provided to link the unit to its destination and source data buffers in the data pools.

The `AocsUnitFunctional` interface is defined, at the design level, as follows:

AocsUnitFunctional
acquireFunctionalData() : void *sendFunctionalData() : void* *refreshFunctionalIn() : void* *refreshFunctionalOut() : void* *ioLinkOperations() : void* *synchronizationOperations() : void* *setFunctionalInConverter() : void* *setFunctionalOutConverter() : void*

The interface exposes methods to acquire and send data, to process them, and to set up links to the unit buffers. In this case, too, transformations between raw and AOCS data levels are performed by control channels that are loaded as plug-in components with methods `setFunctionalInConverter` and `setFunctionalOutConverter`. These two methods would take an argument of type `AbstractControlChannel` representing the control channel that implements the conversion from raw to AOCS data (`setFunctionalInConverter`) and the   control channel that implements the inverse conversion from AOCS data to raw data (`setFunctionalOutConverter`). Typical operations that would be performed on incoming or outgoing data include:

– conversion between raw data and engineering units;
– correction for sensor misalignments;
– bias correction;
– transformation between sensor and satellite reference frame

The use of control channels to encapsulate such operations means that the type and sequence of operations can be easily changed. The processing algorithm in other words is encapsulated in a plug-in component.

It should be noted that both this and the `AocsUnitHousekeeping` interfaces make provisions for only one set of data to be acquired from or sent to the unit because there is only one `acquireData` and `sendData` operation without arguments. Thus, a call to `acquireFunctionalData` must be translated into a set of instructions to acquire *all* the functional data required from the unit in one acquisition cycle.

Consider for instance the case of a sun sensor. Such a unit will normally provide two data words in each acquisition cycle representing the two components of the sun vector as seen by the sensor. The two data words are normally acquired in two distinct low-level acquisition cycles. The interface proposed here would not allow the acquisition of only one sun vector component: a call to `acquireFunctionalData` on the sun sensor component will cause *both* data words to be acquired and to be placed in the data pool locations to which the component is linked.

This may seem like a lack of flexibility. In reality, AOCS units in current use are invariably designed to supply or require only one type of attitude information. This information may physically be constituted of several data words but, conceptually, it represents a coherent whole and users of the AOCS unit are unlikely to ever need only a subset of this information. The proposed interface, on the other hand, has the advantage of simplifying the management of the unit and is suitable for the *fictitious units* described below thus allowing uniform treatment of a large array of components.

15.3 Unit Triggers

Unit proxies are passive components. Data and command exchanges with external units need to be started by some external active component. One solution to this problem would be to leave the task of initiating data exchanges to the components that need the sensor data. Thus, for instance, an attitude controller that uses sun sensor data will be responsible for triggering the sun sensor acquisition. This solution incurs two problems. Firstly, data exchanges may take time and this may force the data consumer unit to poll the device after initiating an exchange. Secondly, data exchanges may have to be done at specific points in the AOCS cycle (typically, sensors are sampled at the beginning of the cycle and actuators are commanded towards the end of the cycle) and these times do not necessarily fall within the activation window of the component that consumes or produces the unit data.

The alternative solution adopted by the unit framelet is to use dedicated components to control data exchanges with AOCS units. Such components are called *unit triggers*. A unit trigger is an active component. Its function is simply to initiate transactions with units that need to be activated at the same time. The unit trigger therefore holds a list of unit proxies and, when it is activated, it goes through the list and directs each item in the list to perform its transactions. The activation code for a unit trigger could be as follows:

```
// Declare lists of units to be triggered
FunctionalInList fInList;
FunctionalOutList fOutList;
HousekeepingInList hkInList;
. . .
void activate() {

    for (all AocsUnitFunctional items u in fInList) do
    { u->acquireFunctionalData();
      u->refreshFunctionalIn();
    }

    for (all AocsUnitFunctional items u in fOutList) do
    { u->refreshFunctionalOut();
```

```
        u->sendFunctionalData();
    }

    for (all AocsUnitHousekeeping items u in hkInList) do
    {  u->acquireHousekeepingData();
       u->refreshHousekeepingIn();
    }
}
```

This pseudo-code assumes that the data acquisition and data send operations are of the blocking kind (they return only after the data transaction has been completed). Other implementations of unit triggers might instead assume non-blocking operations in which case the triggering loop would look like this:

```
    for (all AocsUnitFunctional items u in fInList) do
    {  u->acquireFunctionalData();
       while (!u->isTransactionFinished())
         . . .                          // check for time-out
       u->refreshFunctionalIn();
    }
```

Still other implementations might only do the data acquisition or the data refresh. The AOCS Framework should offer several kinds of unit trigger components covering the following possibilities:

– *Normal Triggers* that perform a full data transfer on the unit including both bus transaction and buffer refresh but they assume that the bus transaction operations are of the blocking kind, ie. that the transaction methods return after the bus transaction has been completed.
– *Polling Triggers* that perform a full data transfer on the unit including both bus transaction and buffer refresh. This means that, on incoming data, they initiate the acquisition transaction, wait for it to be finished and then refresh the acquired data. On the outgoing data, they refresh the outgoing data, initiate the send transaction, and wait for it to be finished.
– *Transaction Triggers* that perform the bus transaction part of the operation associated to the units in the trigger list. This means that they initiate the bus transactions implied by the items in the trigger list but they do not wait for their completion.
– *Refresh Triggers* that perform the buffer refresh part of the operation associated to the units in the trigger list. This means that they call the refresh methods on the unit proxies but do not initiate any bus transaction.

Note that no abstract interface is defined to characterize unit triggers. Unit triggers only exist to be activated and are therefore characterized by the `Runnable` interface that defines active components. Conceptually, unit triggers are similar to functionality managers in the sense of section 8.6. The functionality managers however are defined as core components and only one instance of each would be deployed in an AOCS application. Unit triggers instead exist in several variants

and there may be several instances of unit triggers deployed in the same application.

In a typical configuration, two unit triggers can be used. The first one is scheduled very early in the AOCS cycle and is responsible for triggering data acquisitions from sensors. The second one is scheduled towards the end of the AOCS
control cycle and is responsible for triggering transmissions of data to the actuators. More complex triggering patterns are possible by deploying more unit triggers and by arranging for them to be scheduled at appropriate times in the AOCS
control cycle. This variability gives rise to a hot-spot (the *unit trigger hot-spot*).

Sometimes, the sequence and type of units that have to be triggered vary in different mission phases. Some units for instance might be switched off in some mission phases while others may have to be triggered with different frequencies in
different mission phases. This variability can be handled by endowing the unit
triggers with *operational mode* as discussed in chapter 12. The strategies associated to each operational mode are the lists of units to be triggered. At the beginning
of an activation cycle, the unit trigger then asks a mode manager to supply it with
the unit lists that are valid at that point in time.

15.4 Hardware Unit Components

Direct interaction with the hardware is delegated by AOCS unit proxies to *hardware unit components*. Hardware unit components are characterized by interface
`AocsUnitHardware`:

AocsUnitHardware
powerOperations() : void *sendAndWait(instr : UnitInstruction) : void* *sendAndReturn(instr : UnitInstruction) : void* *resetUnit() : void* *resetTransaction() : void* *isTransactionFinished() : bool* *wasTransactionSuccessful() : bool* *getAddress() : int*

The semantics of the methods defined by this interface are self-explanatory. Essentially, hardware unit components serve as vehicles to send instructions to external units without making any commitment to the specific type of link connec-
ting the units to the AOCS computer. All higher level functions – management of
health status, data conversions, failure reporting, etc – are left to AOCS unit proxies. Hardware units therefore encapsulate the mechanism used to put outgoing
data and commands on the physical link to the unit and to retrieve incoming data
from the same link. The data and instruction to be put on or retrieved from this
link are stored in a structure called `UnitInstruction`. This structure should
provide fields for the definition of the information required to construct a generic
instruction for the hardware interface controlling the exchanges between the

AOCS computer and an external unit. It should be defined to encompass most existing protocols for physical links to external units.

It is important to appreciate that this framelet introduces two levels of abstractions to handle external units. The abstract class `AocsUnit` provides an abstract interface through which individual units can be managed by the AOCS software. The `AocsUnitHardware` interface and the `UnitInstruction` structure are instead intended to provided generic handles through which the hardware interface to external units can be managed by the `AocsUnit` class. Since such hardware interfaces come in many different kinds, it is possible that in some cases the abstract operations declared by `AocsUnitHardware` or the abstract fields offered by `UnitInstruction` are inadequate. In that case, application-specific interfaces have to be defined. However, the existence of the second layer of abstraction – the `AocsUnit` interface – would mean that such change would have no impact on the AOCS framework and only a minimal impact on the AOCS software.

The `AocsUnitHardware` interface gives rise to a framework hot-spot (the *AOCS unit hardware hot-spot*). A developer-friendly framework should predefine as default components the hardware interfaces to the most common links used to control external AOCS units (eg. MACS bus, 1553 bus, serial line, etc.).

15.5 Fictitious AOCS Units

AOCS units have been defined from a syntactic point of view as components that implement two interfaces: `AocsUnitHousekeeping` and `AocsUnitFunctional`. The former interface is used for issuing housekeeping commands to the unit and collecting housekeeping data from it while the latter is used for functional data exchanges with the unit. If the AOCS unit is seen as a server offering certain services, this latter interface captures the behaviour of AOCS units that is visible to the unit's clients. Clients of the unit would normally regard an AOCS unit proxy as an instance of `AocsUnitFunctional` rather than as an instance of `AocsUnitHousekeeping`.

AOCS unit components act as proxies for physical units that exist outside the AOCS computer. Very often, however, a situation arises where a unit-like behaviour is exhibited by components that do not directly represent any external physical equipment. Consider for instance the acquisition of angular rate information from a gyro package. The physical units are gyros that are sometimes individually addressable and to which there correspond AOCS unit proxies in the AOCS software. The users of rate information (e.g. the attitude controllers) will normally interact with a higher-level component that combines the read-out from individual gyros to produce a 3-vector representing the estimated spacecraft angular velocity. From their point of view, this high-level component is a kind of unit on which the same operations can be performed as are normally performed on AOCS units proper. This type of components will be called *fictitious AOCS units*.

Another example of fictitious AOCS unit is a component that combines measurements from a gyro package and from a star sensor and filters them to produce high accuracy attitude estimates. Users of the attitude estimates would like to see this component as a proxy for some high-accuracy attitude sensor.

In order to handle all these situations, the unit framelet proposes the fictitious unit design pattern described in the following sub-section. This design pattern is similar to that used in the ADAGE project [10, 11] to model chains of interconnected unit processing elements.

15.5.1 The Fictitious Unit Design Pattern

This design pattern is introduced to address the problem of combining components that process unit data without impacting the final users of the unit data. The design pattern is based on the concept of fictitious AOCS unit that is defined as a component that implements the `AocsUnitFunctional` interface. A component that implements this interface offers to its clients the same *functional* interface as an AOCS unit and can be treated by them as if it were a unit. The fictitious unit pattern is illustrated by the UML diagram of figure 15.3. It can also be seen as a refinement of the composite pattern.

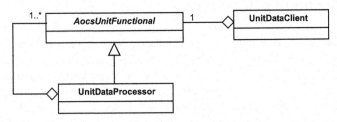

Fig. 15.3. Class diagram for the fictitious unit design pattern

The `UnitDataProcessor` is a concrete class that performs some kind of processing on the unit data. It is a fictitious AOCS unit because it implements interface `AocsUnitFunctional`. The unit data processor obtains the unit data from components that it sees as instances of type `AocsUnitFunctional`. These components may either true AOCS units or fictitious AOCS units. Interaction through the `AocsUnitFunctional` interface shields the unit data processor from having to know whether it is interacting with a real or a fictitious unit. `UnitDataClient` is the final user of the unit data. It too sees its source of unit data as an instance of type `AocsUnitFunctional` with the same advantages.

The fictitious data unit concept allows data processors to be easily combined without disrupting client's operation. An example of a combination of unit data processing elements is shown in figure 15.4 using informal notation. The sensor outputs from two redundant sensors are first processed by a reconfiguration manager that selects the data from the currently active unit. They are then passed through a smoothing filter and are finally received by a controller which is the final user of the sensor data. Its operation is independent of how many filters and other processing elements are interposed between itself and the actual sensors because the interface between successive stages in the processing chain is always the same `AocsUnitFunctional` interface. This makes it easy to insert or remove intermediate processing stages. The implementation of interface `AocsUnit-Functional` to encapsulate the (application-specific) processing of unit data is one of the framework hot-spot (the *fictitious unit hot-spot*).

Note that the fictitious unit design pattern introduces recursion (as does its close relative the composite pattern). Calls to `AocsUnitFunctional` methods may be recursive but the maximum depth of recursion is easily bounded by the maximum number of elements that are linked together in a fictitious unit chain.

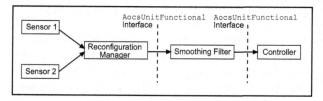

Fig. 15.4: Example of a chain of fictitious units

15.6 Implementation Case Example

Consider the implementation case described in table 26. This section shows how it can be worked out during the framework design definition stage. Its objective is to configure an AOCS unit proxy representing a sun sensor proxy. The main actors in this configuration exercise are:

```
AocsUnit* sunSensorProxy;
AocsUnitHardware* macs;
SensorPool* sensorPool;
AbstractControlChannel* ssDataConverter;
```

The sun sensor proxy would be instantiated from some suitable subclass of `AocsUnit` but is seen by the rest of the AOCS application as an instance of `AocsUnit`. Component `macs` represents the hardware interface to the sun sensor unit. The `sensorPool` component is the data pool holding sensor measurements that was set up in the implementation case of section 13.6. We can assume that the sun sensor produces only two items of data representing the angular direction of the sun vector. Component `ssDataConverter` encapsulates the processing that is required to convert these two measurements from the raw format in which they are acquired from the hardware interface to the AOCS data level format in which they must be deposited into the data pool to be made available to other interested components. The container component for the data processing algorithm is seen as an abstract control channel.

The configuration code then looks like this:

```
// Load hardware interface component
sunSensorProxy->setHardwareInterface(macs);

// Load the data processing algorithms
sunSensorProxy->setFunctionalInConverter(ssDataConverter);

// Link the sun sensor proxy to the data pool locations
AocsData* sunSensor = sensorPool->getSunSensorA();
DataItemWrite destBuffer1 = sunSensor->getDataItemWrite(1);
DataItemWrite destBuffer2 = sunSensor->getDataItemWrite(2);
sunSensorProxy->linkDestBuffer(destBuffer1,1);
sunSensorProxy->linkDestBuffer(destBuffer2,2);
```

The fist statement assumes that class `AocsUnit` defines a `setHardwareInterface` method for loading the hardware interface plug-in component. In the last group of statements, the `AocsData` object where the sun sensor output is located in the data pool is first retrieved from the data pool using a dedicated getter method (see implementation case of section 13.6). The sun sensor `AocsData` consists of two data items representing the two angular coordinates of the sun vector. The sun sensor proxy needs to have write access to them and therefore the `DataItemWrite` objects are recovered from it and are passed to the sun sensor proxy that will use them as the destination buffers for its read-outs.

After the sun sensor is configured as shown here, a call to its `acquireFunctionalData` and `refreshIn` methods will result in the latest sensor read-outs being acquired and deposited in a data pool in AOCS data format ready to be used by other AOCS components.

Table 26. Implementation case example for AOCS unit

Objective	Configure a proxy component for a sun sensor.
Description	A sun sensor proxy is an AOCS unit proxy that represents a sun sensor. This IC shows how such a component can be configured during the application initialization phase.
Framelet	AOCS Unit Framelet
	Intercomponent Communication Framelet
Constructs	AocsUnit abstract class
	AocsData abstract class
	AocsUnitHardware abstract interface
Hot-Spots	None
Related ICs	IC of section 13.6
Implementation	- load hardware interface component into sun sensor proxy.
	- load control channel blocks implementing conversions from raw to AOCS data level into sun sensor proxy.
	- link sun sensor proxy to data pool locations where sun sensor outputs are to be stored.

15.7 Reusability

The design solutions offered by this framelet further reusability within the AOCS domain because they provide a standard interface for AOCS units that decouples the managers and users of unit data from the units themselves. Components can be written in terms of the generic `AocsUnitFunctional` and `AocsUnitHousekeeping` interfaces and can be coupled at run-time with specific units. This is important because the AOCS units and their interfaces are often defined late in an AOCS project. At an early stage, the software designer can tailor the AOCS software to the characteristics of some convenient fictitious units. At a later stage, when the real units are defined, the fictitious unit will act as a transparent bridge between them and the rest of the AOCS software. Furthermore, the fictitious unit concept makes it possible to treat components that process unit inputs and outputs as if they were themselves units thus simplifying the software design.

Table 27. Overview of AOCS unit framelet

Design Patterns	
Fictitious Unit	Allows component that process unit data to be treated in a uniform manner and to be combined without impact on the final users of the data.
Abstract Classes and Interfaces	
`AocsUnitFunctional`	Characterize functional interface of an AOCS unit.
`AocsUnitHousekeeping`	Characterize housekeeping interface of a AOCS unit.
`AocsUnitHardware`	Characterize low-level component that handles interactions with the physical interface to an external unit.
`AocsUnit`	Base abstract class for all components representing external unit proxies in the AOCS software.
Default Components	
Concrete trigger units	Encapsulation of unit trigger profiles.
Concrete hardware interfaces	Encapsulation of common hardware interfaces to external units (MACS interface, 1553 interface, serial line interface, etc).
Hot-Spots	
Aocs Unit	Applications must subclass `AocsUnit` to create proxies for their own units.
Aocs Unit Hardware	Applications must implement `AocsUnitHardware` to model the interactions with the physical links to external units.
Fictitious Unit	Applications must define the unit-like components implementing interface `AocsUnitFunctional`.
Unit Trigger	Applications must define the order in which units are interrogated and the type of triggering action that must be used. This is done by deploying suitable selecting and scheduling unit trigger components.

All embedded systems must manage interfaces to external units but their characteristics vary so much across domains (and, indeed, within a given domain) that it is unlikely that the unit model proposed here could be ported to other domain. The structure of the `AocsUnitFunctional` and `AocsUnitHousekeeping` interfaces in particular reflects the need of AOCS systems (in fact, the introduction of active sensors will soon make them inadequate even in this domain). The problem addressed by the fictitious unit design pattern instead arises in embedded control systems in general and the solution the pattern proposes is not specific to the AOCS and could therefore be ported without changes to other domains.

Table 27 summarizes the constructs exported by the framelet that have been defined or identified in this chapter.

16 The Reconfiguration Management Framelet

Reconfiguration management is one of the essential functionalities of an AOCS made necessary by the high reliability requirements of AOCS systems and by the need to ensure robustness to any single fault. In conventional AOCS applications reconfigurations are performed exclusively on external units. The AOCS Framework proposes a concept of reconfiguration that is wider and more abstract and in principle applicable to any functionality for which multiple implementations are provided. The functionalities that must be reconfigurable and the way in which reconfigurations are performed are obviously application-specific. The problem addressed by the reconfiguration framelet is to define an application-independent mechanism for handling reconfigurations in a generic manner. For this purpose, the framelet offers a dedicated design pattern.

16.1 Some Definitions

If the same functionality can be implemented in two or more independent ways, then the functionality is said to be *redundant*. A redundant functionality can be *reconfigured*. Reconfiguration means switching between different independent implementations of the same functionality.

Reconfigurations usually occur in response to detection of an error: if one implementation of a functionality is found to be faulty, reconfiguration makes the functionality available again by switching from a faulty to a (hopefully) correct implementation.

Functionalities in the AOCS software are implemented as services (method calls) provided by components. The functionality with respect to which reconfiguration takes place is called the *reconfigurable functionality*.

A *reconfiguration group* is a set of components that together offer a redundant functionality. A reconfiguration group can be the object of a reconfiguration. A *redundant component* is a component belonging to a reconfiguration group. The *order* of the reconfiguration group is the number of independent, functionally equivalent, configurations offered by the group. Configurations in a reconfiguration group may be *ranked* according to the performance level with which they implement the group's functionality. A configuration in a reconfiguration group is *marked* either "healthy" or "unhealthy". When a reconfiguration takes place, the configuration the group is configuring away from is marked as "unhealthy".

A. Pasetti: Software Frameworks, LNCS 2231, pp. 209-216, 2002.
© Springer-Verlag Berlin Heidelberg 2002

Some examples are useful to clarify these definitions. A first example of a redundant group is the components representing two aligned sun sensors (prime and redundant sun sensor). In this case, the functionality is the provision of sun position information. A reconfiguration means switching from one to the other sun sensor. The order of this reconfiguration group is 2. If the two sun sensors have identical characteristics, then the two configurations have the same ranking, otherwise the configuration including the most accurate sun sensor should have the higher ranking.

Another example of a redundant group is a set of four non-aligned gyros[19]. The functionality offered by this group is the provision of 3-axis angular rate information. A reconfiguration means to switch from one set of three gyros to a different set of three gyros. The order of this reconfiguration group is four (there are four sub-sets of three gyros in a set of four gyros). The ranking of the four configurations depends on the characteristics of the gyros and on their geometric disposition.

Still another example of reconfiguration group is represented by a set of the following units: a fine sun sensor, a 3-axis gyro package, and a star tracker. The functionality offered by this group is the provision of high-accuracy attitude information. High accuracy attitude information can be provided either by combining fine sun sensor and gyro data or directly by the star tracker. Hence, this group has a reconfiguration order of 2 (in one configuration the sun sensor and the gyro are used, and in the other configuration the star tracker is used). The ranking of the two configurations depends on the relative accuracy of the sensors and on the quality of the data filtering used in the mixed sun sensor/gyro configuration.

16.2 The Reconfiguration Management Design Pattern

This design pattern is introduced to address the problem of separating the management of a reconfiguration group from the provision of the reconfigurable functionality. This pattern is illustrated by the class diagram of figure 16.1.

The redundant components are instantiated from class `Redundant`. The reconfigurable functionality they offer is encapsulated in the abstract interface `ReconfigurableFunctionality`. The reconfigurations are managed by a *Reconfiguration Manager*. The reconfiguration manager controls a reconfiguration group represented by a set of components of type `Redundant`.

The reconfiguration manager is characterized by interface `Reconfigurable` whose key method is `reconfigure`. A call to `reconfigure` triggers a reconfiguration: the reconfiguration manager chooses the highest-ranking configuration among the alternative configurations that are still marked "healthy". The configuration that is abandoned is automatically marked "unhealthy".

---

[19] It is recalled that any three non-aligned gyros can provide full 3-axis rate information.

Components that are responsible for commanding reconfigurations are represented in the class diagram by class `ReconfigurationClient`. They see the
reconfiguration manager as an instance of type `Reconfigurable`. When they
wish to initiate a reconfiguration, they call its `reconfigure` method.

The reconfiguration manager also implements interface `Reconfigurable-`
`Functionality` which makes it "look like" a redundant component. The reconfiguration manager implements the methods declared by `Reconfigu-`
`rableFunctionality` through delegation to the currently active redundant
component in the reconfiguration group. The reconfiguration manager must be
able to act as the sole functional interface between the reconfiguration group and
the users of the reconfigurable functionality. Components that use the reconfigurable functionality (class `FunctionalityClient` in the diagram) see the reconfiguration manager as an instance of type `ReconfigurableFunctiona-`
`lity`.

The implementation of method `reconfigure` in interface `Reconfigu-`
`rable` is one of the framework hot-spot (the *reconfigurable hot-spot*) as it is here
that AOCS applications define their application-specific reconfiguration logic.

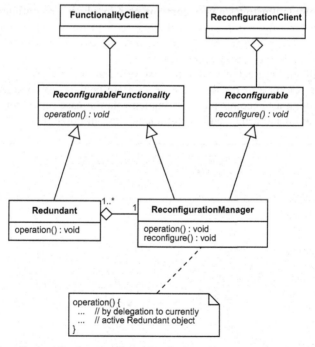

Fig. 16.1. Class diagram for the reconfiguration design pattern

As a concrete example, consider again the gyro reconfiguration group example
described above. The gyros – as AOCS units – would be characterized by interface `AocsFunctional` which then becomes the reconfigurable functionality. A

typical user of this functionality could be an attitude controller that needs 3-axis rate information. A typical component responsible for performing reconfiguration could be a failure recovery component that would trigger a reconfiguration in response to the detection of a gyro failure.

The architecture for this example is shown in figure 16.2 using an informal notation. The lightly shaded boxes represent components. The darker boxes are abstract interfaces. The dashed arrows are implementation links. Thus, gyro 1 is a component that implements interface AocsUnitFunctional which is here symbolically represented by operation acquireData. The solid arrows represent association links with the direction of the arrow representing the direction of the flow of control (e.g. the reconfiguration manager calls acquireData on the gyro components).

The figure shows that the reconfiguration manager has two faces: it has an AocsUnitFunctional face that it exposes towards the attitude controller (to which it supplies the rate estimate obtained from merging the rate measurements from the three active gyros) and it has a Reconfigurable face that it exposes towards the failure recovery manager (to which it supplies a method to reconfigure the set of four gyros to exclude faulty units).

This example illustrates a very common case in which a reconfiguration manager handles reconfiguration across real units. In that case, the reconfigurable interface is AocsUnitFunctional and the reconfiguration manager thus becomes a fictitious unit in the sense of section 15.5.

Reconfiguration are typically either triggered autonomously by the on-board software in response to the detection of a failure or they are commanded by the ground through a telecommand. The implementation case of section 20.4 shows how a telecommand can be built to command a reconfiguration.

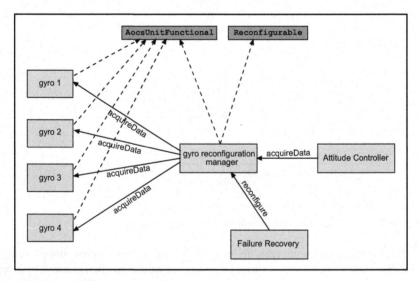

Fig. 16.2. Example of a gyro reconfiguration manager

16.2.1 Intersection and Nesting of Reconfiguration Groups

In a simple implementation, reconfiguration groups are disjoint: if a component belongs to one group, it cannot belong to any other. Disjunction of reconfiguration groups makes their management simple but may not always be optimal. Consider a mission scenario where there are two redundant fine sun sensors (FSS) and only one coarse sun sensor (CSS). The FSSs are intended for a high accuracy mode whereas the CSS is intended for a low-accuracy attitude acquisition mode. Since an FSS can also serve as a CSS (possibly with a smaller field of view), it is conceivable that one of the FSSs should serve as redundant sensor for the CSS. Two reconfiguration groups can then be defined, one providing low-accuracy sun information and the other providing high-accuracy sun information.

Pictorially, this situation is represented in figure 16.3. The second FSS is shared by the two reconfiguration groups. Clearly, reconfiguration information should somehow be passed from one group to the next. Suppose for instance that the FSS reconfiguration manager reconfigures from FSS_2 to FSS_1. This probably means that FSS_2 is faulty and hence the CSS reconfiguration manager should be informed of the fact to exclude it from its reconfigurations.

A simple way to organize this exchange of information would be to endow the FSS with a boolean property called configuredOut. When one of the reconfiguration managers attached to an FSS performs a reconfiguration that excludes the FSS, then the excluded FSS has its configuredOut property set to true. Other reconfiguration managers can then monitor the value of the ConfigureedOut using the change notification mechanism of section 11.3.2. In the case of the figure, the CSS, when notified that FSS_2 has been excluded, could either lower the ranking the configuration that includes it, or it could mark the configurations that include it as "unhealthy".

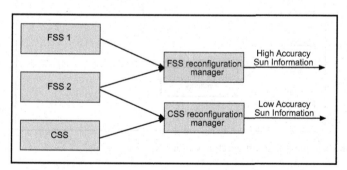

Fig. 16.3. Example of intersection of reconfiguration groups

The general mechanism is for components that straddle reconfiguration groups to carry an observable property configuredOut. This property is set to true when the component is excluded from a configuration. All reconfiguration groups that include the component register their interest in the property and are notified when its value changes. Appropriate action in response to a notification that one component has been excluded from a reconfiguration group could be: lowering the ranking of configurations including that component, or marking configurations including that component as "unhealthy".

Reconfiguration groups can be nested within each other. Consider again the last example described at the beginning of the chapter. Attitude information can be derived either by combining fine sun sensor (FSS) and gyro (GYR) data or from a star tracker (STR). If it is assumed that the individual sensors are also redundant, the structure of figure 16.4 results. In this case, the components reconfigured by the attitude reconfiguration manager are themselves reconfiguration managers. The reconfiguration order of the attitude reconfiguration manager is determined by the reconfiguration orders of the FSS, GYR and STR reconfiguration orders.

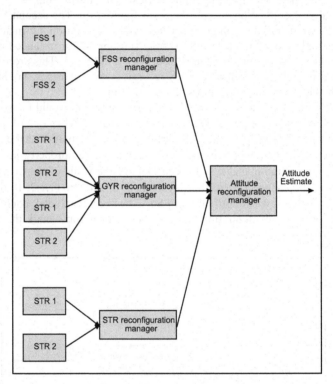

Fig. 16.4. Example of Reconfiguration Groups Nesting

16.2.2 Direct Access to Redundant Components

Is direct access to redundant components required? For instance, in the case of the four gyros arranged as in figure 16.2, is direct access to the gyro components ever required or should access to the gyros always be through the reconfiguration manager? Two types of access to redundant components can be envisaged: functional access and housekeeping access. Redundant components provide services, *functional access* is the access by a client that needs these services. In the case of a gyro, for instance, a functional access is an access to obtain rate estimates. *Housekeeping access* is aimed at performing housekeeping operations on the component (initialization, mode changes, self-test, etc.). Functional access is exclusively through the reconfiguration manager. For this reason, the reconfiguration manager is made to implement the `ReconfigurableFunctionality` interface that characterizes the redundant components from a functional point of view.

Housekeeping access should preferably be done through the reconfiguration interface but can occur directly when needed. The reconfiguration manager may have to be kept informed of housekeeping operations performed on its redundant components. In this case, the monitoring mechanism of section 11.3.2 is used. Thus, for instance, if housekeeping access to a gyro is allowed to change the gyro's health status, then this health status is treated as a property with which the gyro reconfiguration manager can register.

16.2.3 Preservation of Configuration Data

A reconfiguration is usually the result of a detected or suspected fault in a component that, through the reconfiguration, is excluded from the normal flow of AOCS data. Reconfiguration information should therefore be preserved across software and hardware resets in order to allow safe autonomous re-initialization of the AOCS software. The storage of configuration information data follows the memento design pattern. Configuration information is stored in *configuration state* objects. Such objects encapsulate the configuration of a reconfiguration group. In particular, they store the following information:

- Configurations that have been marked "unhealthy"
- The current configuration

All reconfiguration managers can give out a configuration state object that holds their current configuration in a format that is specific to the reconfiguration manager itself. Responsibility for preserving configuration information in the event of a software or hardware reset lies with the system manager of chapter 10. Reconfiguration managers notify the system manager whenever their internal configuration has changed and give it a copy of their configuration state object describing their new configuration. After a reset, the system manager will hand back the latest configuration state objects to the reconfiguration managers who will then use them to restore their pre-reset configuration.

16.3 Reusability

The design proposed in this section furthers reusability because it decouples the task of managing reconfigurations from the handling of the operational interactions with the components that are reconfigured. Clients of the reconfigured components always see the same interface to represent a group of reconfigurable components, regardless of which particular component is selected at a particular point in time. The management of the reconfigurations of a group of components is in turn independent of who the component clients are and of how they interact with the component itself.

The reconfiguration framelet can be used in contexts other than the AOCS because it defines reconfigurations with respect to an abstract functionality as encapsulated by an abstract interface. Reuse is possible at both the design and architectural level wherever there is a need for managing multiple implementations of the same functionality.

Table 28 summarizes the constructs exported by the framelet that have been defined or identified in this chapter.

Table 28. Overview of reconfiguration framelet

Design Patterns	
Reconfiguration Management	Separate the management of reconfigurations from the implementation of reconfiguration logic and of the reconfigurable functionality.
Abstract Interfaces	
`Reconfigurable`	Interface that characterizes a reconfiguration manager, namely a component that can perform reconfigurations.
Core Components	
Configuration state	Memento object holding the configuration data for a reconfiguration manager.
Hot-Spots	
Reconfigurable	Applications must define the reconfiguration logic for their reconfiguration managers by providing implementations for interface `Reconfigurable`.

17 The Manoeuvre Management Framelet

The term *manoeuvre* designates a sequence of actions that must be performed by the AOCS at specified times to achieve a specified goal. Although individual manoeuvres – to unload wheel momentum, to slew the spacecraft, etc – exist in almost any AOCS system, their diversity is such that usually no concept of abstract manoeuvre is used. From the point of view of a framework, the actions to be performed by a manoeuvre constitute a hot-spot and the primary task of this framelet is to model the associated variability. The solution it offers is built upon an abstract interface that defines the abstract operations that can be performed upon manoeuvres. The implementation of this interface in concrete manoeuvre components is left to the application developers (although some default components encapsulating commonly recurring manoeuvres are provided by the framework). Additionally, a core component – the *manoeuvre manager* – is provided that is responsible for controlling the execution of manoeuvres. This is one of the functionality managers discussed in section 8.5. The interaction between the manoeuvre manager and the abstract manoeuvres is defined by a dedicated design pattern that separates the management of the manoeuvres from their implementation.

17.1 Manoeuvre Components

In order to allow uniform and mission-independent treatment of manoeuvres, it is necessary to encapsulate them in components derived from a common base class or implementing the same abstract interface. The framelet defines the following abstract interface to characterize manoeuvre components:

Manoeuvre
canStart() : bool
canContinue() : bool
isFinished() : bool
doContinue() : void

Manoeuvres execute over a time interval which may be defined by the manoeuvre itself. Method `canStart` returns `true` when the manoeuvre is ready to start. Depending on the manoeuvre, readiness to start can be defined by a time tag or by the occurrence of certain operational conditions. An example of the latter is the case of a manoeuvre to unload the angular momentum in the reaction wheel (see section 2.1.1). Its execution would typically be triggered by the speed of the

A. Pasetti: Software Frameworks, LNCS 2231, pp. 217-221, 2002.
© Springer-Verlag Berlin Heidelberg 2002

reaction wheel exceeding a pre-defined threshold. Method `canStart` in such case would implement a check on the reaction wheel speed and would return true if the speed is found to be above the threshold.

Manoeuvres are executed in several steps. Method `doContinue` is called by the manoeuvre manager to advance the execution of the manoeuvre. When a manoeuvre component receives this command, it check the current time or it verifies current operational conditions and it performs any actions that are due for execution. Manoeuvres are thus endowed with the ability to check whether the conditions for their own continued execution are appropriate. For instance, a manoeuvre to perform an attitude slew should periodically check that the spacecraft is following the slew profile. If the deviation from the slew profile exceeds some pre-specified threshold, then an error has occurred and the manoeuvre should be aborted. In another example, consider the manoeuvre to perform an open-loop delta-V. This manoeuvre should only be executed if thruster operation is enabled. If this is not the case, the manoeuvre should be aborted. Method `canContinue` can be called to verify if the conditions for the continued execution of the manoeuvre hold.

Method `isFinished` returns `true` when the manoeuvre has terminated. Termination can be defined by a time tag or by the achievement of pre-specified operational conditions. In the case of an attitude slew, for instance, the termination of the manoeuvre is determined by the spacecraft having reached its target attitude.

Since manoeuvres are application-specific, the implementation of the abstract class `Manoeuvre` represents one of the framework hot-spots (the *manoeuvre hot-spot*). Although manoeuvres are application-specific, there are some kinds of manoeuvres that recur often. This is the case of attitude slews, execution of delta-V and unloading of reaction wheels. A mature framework should provide default components that implement standard versions of these manoeuvres that can be configured for use in a particular application.

It might even make sense to define base classes covering groups of related manoeuvres. A very common kind of manoeuvre in an AOCS consists in driving a controller to follow a pre-defined profile. This will typically result in the spacecraft performing slews but profiles can also be used to perform wheel unloading (in which case it is the wheel speed that is controlled in closed loop to follow a pre-defined profile). Given the frequency and importance of this type of manoeuvre, a dedicated superclass for them that extends the base class `Manoeuvre` could be defined thus giving rise to a class tree for the AOCS manoeuvres.

17.2 The Manoeuvre Design Pattern

This design pattern is introduced to address the problem of separating the management of manoeuvres from their implementation. The design pattern is based on the manager meta-pattern of section 8.5 and is illustrated in the class diagram of figure 17.1.

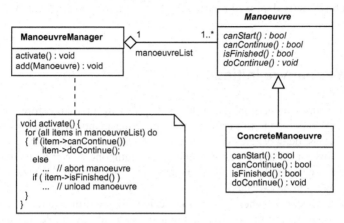

Fig. 17.1. Class diagram of the manoeuvre manager design pattern

The manoeuvre manager holds a list of manoeuvres that are seen as instance of interface `Manoeuvre`. Its presence separates the management of manoeuvres from their implementation.

At each activation, the manoeuvre manager goes through the list of pending manoeuvres, checks which ones are due for execution and which ones are already executing and are in a condition to continue execution and on the latter it calls method `doContinue` to advance the manoeuvre execution. Finally, the manoeuvre manager checks whether manoeuvres have terminated their execution and, if they have, removes them from the list of pending manoeuvres.

Note that, in principle, clients can only *add* manoeuvres to the manoeuvre manager list. Removal from the list is done autonomously and internally to the manoeuvre manager when a manoeuvre has terminated its execution.

The manoeuvre pattern is instantiated within the framework and should not be directly visible to the framework users (the application developers). The first issue in instantiating this design pattern concerns the activation of the manoeuvre manager. In the AOCS Framework, the manoeuvre manager is implemented as an active component and its `activate` method is the `run` method declared by interface `Runnable`.

AOCS applications usually have a requirement that the execution of manoeuvres be logged. In accordance with the concepts proposed in chapter 13, logging is done using events. A dedicated manoeuvre event class is defined and the manoeuvre manager records manoeuvre-related events as instances of this class (see the implementation case of section 13.5 for an illustration of how this could be done).

Additionally, methods are provided to allow on-going manoeuvres to be aborted in an orderly manner. Manoeuvre abortion could be the result of a ground command or could be decided autonomously on-board in response to failures detected in other parts of the software.

The manoeuvre manager – like the other functionality managers in the AOCS Framework – is application independent and can be packaged in a core component that can be directly deployed in any AOCS application to be instantiated from the framework.

17.3 Manoeuvre Initiation

In a typical implementation, the AOCS software will internally store a number of manoeuvre components to perform often recurring manoeuvre such as wheel unloading, slews, delta-V, etc. When conditions warrant it or when the ground commands it, one of these pre-defined manoeuvre components can be loaded into the manoeuvre manager (by calling its method add(Manoeuvre)). This will cause the manoeuvre to be executed.

Thus, for instance, when the ground decides to perform a delta-V manoeuvre, and assuming that the manoeuvre component is already available on board, it can uplink a telecommand that performs the following method call:

```
manoeuvreManager->add(deltaVManoeuvre);
```

where deltaVManoeuvre is the component that encapsulates the delta-V manoeuvre. In another example, if the failure detection mechanism has detected an over-speed condition in one of the reaction wheel, it will report the fact as an event. This event will then be processed by the failure recovery manager that could respond by loading the reaction wheel manoeuvre into the manoeuvre manager.

17.4 Alternative Solution

There is an obvious similarity between manoeuvres as defined here and telecommands as defined in the telecommand framelet of chapter 20. Both categories of components encapsulate actions that must be performed on the AOCS software. Perhaps the main conceptual difference is that telecommands are intended to be performed in one shot whereas the execution of manoeuvres extends over time. This is not an essential difference though, because telecommands could be seen as special cases of manoeuvres with a punctual execution (their isFinished method always returns true). In the AOCS Framework, manoeuvres and telecommands are kept separate primarily because they are normally seen as distinct by the AOCS designers but also for implementation reasons. Telecommands have special implementation needs because, unlike other components in an AOCS application, they must be created dynamically upon reception of messages from the ground. This requirement is likely to have an impact on the management of the telecommands and advises against treating them as a special kind of manoeuvres.

17.5 Reusability

Reusability is achieved by decoupling the task of *managing* manoeuvres from the task of *carrying them out*. In this way, the manoeuvre manager becomes mission-independent and reusable across missions without changes. Manoeuvre are encapsulated in components and can therefore be changed with only local repercussions thus again promoting reusability.

The concept of manoeuvre is applicable in other embedded domains and could be reused elsewhere at design level. Reuse at architectural level might also be possible in domains where the semantic of manoeuvres is the same as in the AOCS domain. Reuse at code level would probably only involve the manoeuvre manager and be restricted to domains where the logic to handle manoeuvres is the same as in the AOCS domain.

Table 29 summarizes the constructs exported by the framelet that have been defined or identified in this chapter.

Table 29. Overview of manoeuvre framelet

Design Patterns	
Manoeuvre Management	Separate the management of manoeuvre execution from the implementation of the manoeuvres.
Abstract Interfaces	
Manoeuvre	Interface characterizing manoeuvre components.
Core Components	
ManouvreManager	Component to manage the execution of manoeuvres.
Default Components	
Recurring Manoeuvres	Components encapsulating commonly used manoeuvres (reaction wheel unloads, delta-V, attitude slews, etc)
Hot-Spots	
Manoeuvre	Applications must define their own manoeuvres by subclassing Manoeuvre.

18 The Failure Detection Management Framelet

The term *failure* is used in the framework to designate any fault that is detected during the normal operation of a component and that affects its ability to perform its allotted task. Failures occur during normal operation and have their origin in the environment around the software. A typical example is a malfunction in an external unit that causes inconsistent data to be fed to the AOCS software. Another example is a controller instability following operation outside nominal conditions due, for instance, to unusually high satellite angular velocity.

The general policy in the AOCS domain is that the system should be robust to any single failure meaning that full performance should be maintained in the presence of any single failure. This normally implies that the software should be capable of detecting failures and of responding to them in a manner that completely removes their consequences (provided that only one failure has occurred).

Failure detection is part of a more general function that in the AOCS domain is often referred to as FDIR standing for "Failure Detection, Isolation and Recovery". The FDIR is often a critical part of an AOCS software both in the sense of being essential to meeting mission objectives (in particular with regard to robustness to single failures) and in the sense of being one of the most complex and error-prone. The FDIR algorithms are very mission-specific and cannot therefore be part of a generic AOCS Framework. Hence, the main issue in designing the framework is *not* the identification and implementation of generic FDIR algorithms but it is rather the definition of an architecture that will support the implementation of such algorithms that are then seen as major framework hot-spots.

The failure detection framelet proposes a solution to the problem of managing failure detection checks. This solution is based on a dedicated design pattern derived from the manager meta-pattern. At its centre stands the *failure detection manager* component that is responsible for initiating failure checks and for creating failure reports in case failures are detected. The failure detection manager is one of the framework functionality managers and is fully reusable across AOCS applications.

18.1 Overall Approach

Failure handling in the AOCS Framework is based on three principles. Firstly, failure detection and failure recovery are regarded as separate functions allocated to distinct components. The failure detection manager performs failure checks

A. Pasetti: Software Frameworks, LNCS 2231, pp. 223-231, 2002.

and, when it detects a failure, it creates a failure report that describes it and stores it in some repository. Other components are responsible for inspecting this repository to check whether any failures have been identified and then use the information stored in the failure reports to decide on the appropriate response. The failure detection framelet thus only covers the failure detection part of the FDIR function. Failure recovery is covered by a second framelet discussed in the next chapter. The problems related to the isolation of failures are briefly considered at the end of this chapter.

Secondly, the knowledge required to perform a failure detection check must be confined to the component that is being checked. The task of the failure detection manager is simply to ask for the check to be performed, to record the results and to pass them on to other interested components. The actual check must be performed by the tested component upon itself.

Finally, to each failure detection check there must correspond one or more recovery responses that can be invoked to recover from that failure. Thus, whenever the application designers incorporate a component that performs a failure check into an application, they must also specify the corrective actions to be associated to the detection of that failure. This is necessary because there is no point in detecting failures for which no remedial action can be identified. Responses to failures are themselves encapsulated in dedicated components that are called *recovery actions* (see section 19.1) and that represent sets of related actions that address a specific failure.

18.2 Failure Detection Checks

Failure detection checks performed in AOCS systems normally fall under one of two headings: *consistency checks* and *monitoring of property values*. The failure detection framelet offers mechanisms to support the execution of both.

18.2.1 Consistency Checks

A consistency check verifies that the internal state of a component is consistent. For instance, a consistency check on a quaternion verifies that the squared sum of the quaternion components is equal to 1 (within a certain tolerance band). In another example, a consistency check on a sensor verifies that the sensor's outputs are in the range of physically possible values. In accordance with the second principle mentioned above, an interface is defined by the framelet to characterize components that can perform a consistency check upon themselves:

ConsistencyCheckable
doConsistencyCheck() : bool *setRecoveryAction(RecoveryAction) : void* *getRecoveryAction() : RecoveryAction*

This interface must be implemented by all components that are capable of performing a consistency check on their internal state. A call to doConsistency-Check causes the consistency check to be executed. When the consistency check succeeds (i.e. when no errors are found), the method returns true. If the check finds an inconsistency in the component's state, it returns false and a failure report is then prepared to describe it. The report contains all the information that the component can make available to assist the identification of the cause of the failure. This report is stored in a repository from where it can be retrieved by other framework components (typically, the failure recovery manager, see the next chapter).

The doConsistency method is one of the framework hot-spots (the *consistency checkable hot-spot)* because it is the point where applications define their application-specific consistency checks.

To each failure check must be associated a recovery action. Given the diversity of recovery actions, the simplest option is to assume that they are encapsulated in objects of some generic type RecoveryAction probably representing an abstract base class or an abstract interface. The ConsistencyCheckable interface provides setter and getter methods to allow the recovery action component to be defined and retrieved.

At implementation level, the performance of consistency checks is delegated to a failure detection manager component. This component internally maintains a list of components that must be subjected to the consistency check. When it is activated, it executes code like:

```
ConsistencyCheckableList* list;

for (all objects cc in list) do {
    if ( !cc.doConsistencyCheck() )
    { RecoveryAction* r = cc->getRecoveryAction();
        . . . // failure detected! Form failure event,
        . . . // attach recovery action r to it, and store
        . . . // it in its event repository
    }
}
```

Thus, at each activation, the detection manager cycles through the components in its list and calls doConsistencyCheck on each. The reporting of failures is done through the event mechanism proposed by the intercomponent communication framelet. This means that, whenever it finds that a consistency check has failed, the failure detection manager creates a failure event and stores it in the failure event repository where it is accessible to other framework components.

18.2.2 Property Monitoring

The concept of property monitoring was introduced in chapter 11. The monitored property is encapsulated in an object of type Property. The profile against which monitoring is to be performed is encapsulated in an object of type ChangeObject. The failure detection framelet defines an additional object, the *monitoring check object*, that encapsulates a failure detection check based on property monitoring. A monitoring check object packages in a single object: the property to be checked; the change profile against which the property is monitored; the recovery action to be performed in case the change is found to have occurred. A monitoring check object is an object that is instantiated from the following class:

MonitoringCheck
setProperty(Property) : void
getProperty() : Property
setChangeObject(ChangeObject) : void
getChangeObject() : ChangeObject
setRecoveryAction(RecoveryAction) : void
getRecoveryAction() : RecoveryAction

The implementation of the object monitoring part of failure detection is done by maintaining a list of monitoring check objects and passing each property value through the corresponding change object. The corresponding pseudo-code is:

```
MonitoringCheckList* list;

for (all monitoring check objects mc in list) do {
    Property* p = mc->getProperty();
    ChangeObject* c = mc->getChangeObject();
    RecoveryAction r = mc->getRecoveryAction();
    if ( c->checkValue(p->get()) )
        . . . // failure detected! Form failure event,
        . . . // attach recovery action r to it, and store
        . . . // it in its event repository
}
```

When a change is detected, the fact is reported as an event stored in the failure event repository. The event also records the recovery action associated to the property and its change objects. Note how the checking code is independent of which variable is being checked and of which profile it is being checked against.

Monitoring checks have the following characteristics:

– Their memory and execution overheads are very small
– They are localized: they can be used to perform checks on individual variables or on small groups of variables
– They perform checks that are independent of the algorithms implemented by the components whose behaviour they help monitor
– They can be dynamically and selectively enabled and disabled
– They can be executed continuously and in parallel to normal system operation

In view of these characteristics, monitoring checks can be seen as an implementation of checker technology [15]. Checkers are well-suited to systems like the

AOCS that are mission-critical and where real-time behaviour and intrinsic complexity give rise to error situations that are difficult to reproduce [83, 84].

The type of monitoring check is obviously application specific and therefore the instantiation of the `MonitoringCheck` class is one of the hot spots of the framework (the *monitoring check hot-spot*).

18.3 The Failure Detection Design Pattern

This design pattern is introduced to address the problem of separating the management of failure detection tests from their implementations. It is based on the manager meta-pattern of section 8.5. It is illustrated in the class diagram of figure 18.1.

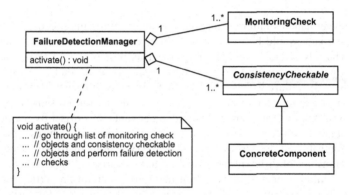

Fig. 18.1. Class diagram of failure detection design pattern

The failure detection manager maintains lists of monitoring check objects and of consistency checkable components. When it is activated, it goes through the lists and performs the failure detection tests on each item in the lists. When a failure is detected, a failure report is created to describe it and it is stored in a repository from where it is accessible to other components in the framework.

This pattern is instantiated in the framework and is normally not visible to framework users. Its instantiation is performed by turning the failure detection manager into an active component. Its `activate` method is then the `run` method declared by interface `Runnable`.

As already mentioned, the implementation of the failure reports to be generated by the failure detection manager when it encounters a failure is done through the event mechanism of the intercomponent communication framelet. A dedicated class of events is created – the failure events – and its associated event repository is used to make the failure reports available throughout an AOCS application.

In most cases, the list of components to be subjected to consistency checks and the list of monitoring check objects depends on operational conditions. This is taken into account by making the failure detection manager mode-dependent. The failure detection mode manager then manages two strategies corresponding to the list of consistency checkable components and to the list of monitoring check objects.

The failure detection manager implements systematic failure checks but other components can report failures that they encounter during their normal operation by creating the corresponding failure event and storing it in the failure event repository. Consider, for instance, the case of an AOCS unit component that tries to access the AOCS bus to send an instruction to an external unit and finds that a bus error is returned. This component has detected an error. It must respond to the detection by creating an event and storing it in the repository for failure events.

18.4 Alternative Approaches

This framelet handles failure detection separately from failure recovery. The alternative approach would be to perform failure recovery as soon as a failure is detected. The advantage of this latter approach is promptness of response and the potentially greater amount of information about the failure that is available at the point where the failure is detected. The first advantage however is minimal – provided that the failure recovery manager is scheduled with the same frequency as the failure detection component. The second advantage is offset by the possibility that a separate failure recovery manager has to inspect *all* failures detected in a certain control cycle to perform global recovery actions that take account of the interrelationships of different individual failures reports.

Part of failure detection is based on monitoring certain properties for specific types of change that might indicate that a failure has arisen. Monitoring can be done either directly or by change notification. This framelet bases failure detection on the direct monitoring mechanism. However, monitoring by change notification would also have been possible. In that case, the failure detection manager would have a purely passive role: it would simply provide a `PropertyChange` method to be called by the monitored component whenever a change has been detected. The difficulty here is that the failure detection manager then becomes a listener to a potentially large number of very similar events and this means that it would need to implement some logic to filter these events and to associate the correct handler to each. The selected approach seems simpler and minimizes the coupling between the failure detection manager and the components that are being subjected to failure detection checking.

18.5 Failure Isolation

The detection of a failure should ideally be followed by the isolation of its cause. Failure isolation is not supported by the AOCS Framework because no generic failure isolation mechanism could be identified. It is however interesting to briefly describe a candidate mechanism based on the concept of *failure tracing* that was considered but eventually discarded due to its excessive complexity.

When a consistency or monitoring check fails on a component, the cause of the failure may lie either internally to the component itself or may be the result of a failure propagation from some other component. In general, a component has one or more *source components* (ie the components from which its data originate) and one or more *sink components* (ie the component where its data are forwarded). One way to support failure isolation is therefore to endow components with knowledge about who their sources and sinks are.

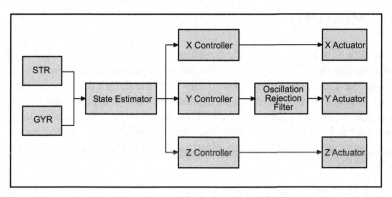

Fig. 18.2. Example of failure tracing through a chain of interconnected components

Consider for instance figure 18.2 that illustrates the conceptual data flow[20] for a controller that uses star tracker and gyro outputs filtered by a state estimator to compute commands for three actuators. If a consistency check reveals a failure in the XActuator component, then it is useful for the failure isolation manager to be able to check whether its data sources also fail their consistency check. This might indicate that the source of the failure is not in the XActuator component but in some other component upstream. Similarly, if a consistency check fails on component StateEstimator, it may be useful to check downstream components. If all downstream components also fail their consistency check then this strengthens the hypothesis that the StateEstimator (or one of its data sources) are the origins of the failure. In order to support this type of analysis, an interface of the following kind could be imposed on all framework components:

DataFlowObject
getDataSource() : DataFlowObjectList *getDataSink() : DataFlowObjectList*

---

[20] This is only a *conceptual* data flow because, as explained in chapter 13, the actual data are not transmitted directly from one object to another but are exchanged through shareable data areas (the data pools).

Components implementing this interface are part of a data flow path. Methods `getDataSource` and `getDataSink` return the list of data source and data sinks for the component and therefore allow tracing the data path both upstream and downstream. This mechanism was not implemented in the AOCS Framework partly because failure isolation is not normally performed in existing AOCS applications and partly because of the complexity of the algorithms required to process the tracing information to arrive at the identification of a single faulty component.

18.6 Reusability

Reusability with respect to failure detection is achieved within the AOCS domain by decoupling the task of managing the failure detection function from the task of carrying out failure detection tests. In this way, the failure detection manager becomes reusable across applications because all AOCS applications perform the same type of failure checks. Failure detection tests are encapsulated in components and can be changed with only local repercussions. Changing the components that are subjected to failure detection testing can be done dynamically without any impact on the software architecture.

Table 30. Overview of failure detection framelet

Design Patterns	
Failure Detection Management	Separate the management of consistency and monitoring checks from their implementation.
Abstract Classes	
`ConsistencyCheckable`	Characterize components that can be asked to check the consistency of their internal state.
Core Components	
`FailureDetectionManager`	Component to manage the execution of consistency and monitoring checks.
`MonitoringCheck`	Component encapsulating a monitoring check as consisting of the property to be checked, the change object representing the profile against which the property is to be checked, and the recovery action associated to the failure detection check.
Hot-Spots	
Consistency Checkable	Applications must define their consistency checks by implementing interface `ConsistencyCheckable`.
Monitoring Check	Applications must define the properties to be monitored, their nominal behaviour and the recovery action to be associated to a failure of the monitoring check. This is done by instantiating `MonitoringCheck` objects.

As usual, the rigid decoupling of the management from the implementation of a function opens the way to the design solutions encapsulated in the framelet being exported to other domains. However, the framelet assumes that failure detection tests are always reducible to consistency and monitoring checks and while these two categories are very wide they may not be suitable for all domains. Reusability at architectural and code levels is only possible where failure detection is performed according to the same logic as in the AOCS domain and using the same identification of failure reports with failure events.

Table 30 summarizes the constructs exported by the framelet that have been defined or identified in this chapter.

19 The Failure Recovery Management Framelet

There is usually no standardized approach to failure recovery in present AOCS applications. Responses to a failure are coded in dedicated procedures that are called at the point where the failure is detected. This framelet proposes a more general solution by separating the *management* of the failure recovery process from the *implementation* of specific failure recovery counter-measures. The latter are obviously application-specific and therefore are treated as a framework hot-spot. The failure recovery function is implemented by a *failure recovery manager* based on the manager meta-pattern. The failure detection manager is one of the framework functionality managers. Its job is to react to failure reports (as they would be typically generated by the failure detection function) by triggering the execution of failure recovery strategies.

19.1 Failure Recovery Actions

In general, the AOCS software can react to a failure report by performing one or more *failure recovery actions*. A failure recovery action represents a *local* response to a failure. The response is said to be local because it is based on a single failure report. A *global* response would take account of sets of failure reports.

Failure recovery actions vary across applications. As is usual in the AOCS Framework, this type of variability is addressed by introducing an abstract base class or abstract interface. This framelet defines the following abstract base class to characterize a generic recovery action:

RecoveryAction
doRecovery() : void   setNextRecoveryAction(RecoveryAction) : void

The key method is `doRecovery` that causes the recovery action to be performed. Its implementation encapsulates the application-specific response to failures and represents a framework hot-spot (the *recovery action hot-spot*).

The same failure report may give rise to several failure recovery actions. For this purpose, recovery actions can be chained together. Method `setNextAction` is used to build up a chain of recovery actions.

The implementation of the recovery action concept in the AOCS Framework identifies failure reports with failure events. This is in accordance with the model

A. Pasetti: Software Frameworks, LNCS 2231, pp. 233-242, 2002.
© Springer-Verlag Berlin Heidelberg 2002

for data exchanges prescribed by the intercomponent communication framelet. Typical failure recovery actions in use in the AOCS domain include:

- *Reset of the entire AOCS software*: This recovery action uses the system reset service of section 10.2 to perform a complete software reset of the AOCS.
- *Reset of one or more AOCS components*: All AOCS components must implement a reset method. This allows the failure recovery manager to selectively reset some components.
- *Reconfiguration of one or more units*: Reconfigurable units are gathered together in reconfiguration groups for which a reconfiguration manager is responsible (see section 16.2). Reconfigurations are performed by calling the `reconfigure` method on the reconfiguration manager.
- *Fall-back to a lower operational mode*: In the AOCS Framework, operational mode is a local property of each component. Additionally, a mission mode manager (see section 12.4) is present to expose the AOCS mission mode. Mode fall-backs can therefore be implemented either at local level (changing the mode of a single component) or at AOCS level (changing the mode of the mission mode manager).

Since the above are very common recovery actions, the framework offers them as *default components* implemented as subclasses of `RecoveryAction`. Developers can of course build additional application-specific recovery actions.

It is worth noting that recovery actions can be endowed with "memory". Consider for instance the case of a Kalman Filter to which a recovery action is associated that is triggered when the filter diverges. The nominal recovery for a filter divergence may be a filter reset. However, the recovery action may be made to remember the last time it was called and, if it finds that it is called too frequently, it can decide that there is a fundamental control failure and may react by commanding a mode fall-back or a complete system reset.

19.2 Failure Recovery Strategy

A *failure recovery strategy* is a set of coordinated responses to all the failure reports generated over a certain time interval. Unlike recovery actions, that act on individual failure reports, recovery strategies can take into consideration several failure reports and can therefore offer a higher-level response to failures. Failure strategies are derived from the following base abstract class:

RecoveryStrategy
doRecovery() : void setNextRecoveryStrategy(RecoveryStrategy) : void

Method `doRecovery` inspects the set of failure reports and implements the appropriate failure recovery response. Its implementation represents the application-specific response to the failures and represents a framework hot-spot (the *recovery strategy hot-spot*).

Failure strategies can be chained together. This allows several failure strategies to be implemented in sequence. Method `setNextRecoveryStrategy` is used to link failure strategies in a chain.

In the AOCS Framework, the concept of recovery strategy is implemented by identifying the set of failure reports to be analyzed by a recovery strategy with the set of failure events in the failure event repository. This is in accordance with the model for data exchanges prescribed by the intercomponent communication framelet.

Concrete failure recovery strategies are implemented as instances of concrete subclasses of `RecoveryStrategy`. Typical failure recovery strategies include:

- *Sequence of local recovery actions*: This strategy retrieves failure events from the event repository in sequence and performs the recovery action associated to each event.
- *System reset on too many failures*: This strategy checks the number of failure events in the repository and if it finds that it exceeds a predefined threshold, it assumes that there is some serious problem and commands a system reset.

The framework offers default components that encapsulate the above recovery strategies. Other more application-specific recovery strategies can be developed on an *ad hoc* basis by subclassing `RecoveryStrategy`.

19.3 Failure Recovery Design Pattern

This design pattern is introduced to address the problem of separating the management of failure recovery from the implementation of failure recovery actions and strategies. The design pattern is illustrated in the UML diagram of figure 19.1.

The failure recovery manager is modelled on the chain of responsibility design pattern but it can also be seen as a refinement of the manager meta-pattern where the list of functionality implementers only contains one element.

In the classical version of the chain of responsibility pattern, the client's request (in this case, the request to perform a recovery) is passed along the chain of handlers (the recovery strategies) until one is found which is able to handle it. Each request is intended to be handled by only one handler. In the application of the pattern to failure recovery, however, a recovery strategy when it receives a recovery requests performs the following actions:

- it handles the recovery request, and
- it checks whether the recovery request should be passed on to the next recovery strategy or whether recovery processing should terminate.

The recovery strategies are executed in sequence but every recovery strategy has the chance to interrupt the chain. This incidentally means that the *order* in which the recovery strategies are linked in the list is important. It would have been possible to implement failure recovery using a more straightforward version of the manager meta-pattern where the recovery strategies are arranged in a list and the recovery manager, when it is activated, goes through the list and executes each strategy in sequence. This solution, however, would have made it more awkward to give each recovery strategy the option to interrupt the recovery process.

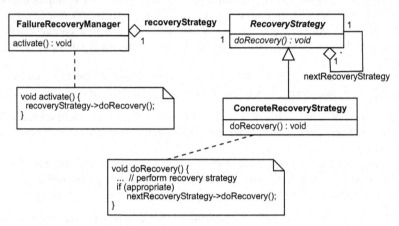

Fig. 19.1. Class diagram of the failure recovery design pattern

The failure recovery pattern is an instantiated pattern in the sense of section 3.2, namely it is a pattern that is used within the framework but should not be visible to the users of the framework who only deal with individual recovery actions and recovery strategies. The failure recovery manager is implemented as an active component and its `activate` method is identified with the `run` method declared by interface `Runnable`. A response to a failure is an important event in an AOCS application and there is normally a requirement that it be logged. In the spirit of the intercomponent communication framelet, this requirement is implemented by creating a class of recovery events and endowing the failure recovery manager with the ability to create an event whenever it executes a recovery strategy or a recovery action.

In most cases, the recovery strategy to be executed depends on operational conditions. This is taken into account by implementing the failure recovery manager as a mode-dependent component using the design pattern of chapter 12. A mode manager is then associated to the failure recovery manager. The mode manager controls a single strategy corresponding to the recovery strategy to be supplied to the recovery manager. This is illustrated in the implementation case example below.

Use of the chain of responsibility design pattern introduces the possibility of recursion. A call to `RecoveryStrategy::doRecovery` can be recursive if several recovery strategies are linked together. The maximum depth of the recursion is given by the maximum number of recovery strategies that are linked together. Recursion can also arise because of the way recovery actions are linked together. A call to method `RecoveryAction::doRecovery` can be recursive if several recovery actions are linked together. The maximum depth of the recursion is given by the maximum number of recovery actions that are linked together.

19.4 Implementation Case Example – 1

Consider the implementation case described in table 31. This section shows how it can be worked out during the framework design definition stage. Its objective is to create and configure a mode manager for the failure recovery manager. The failure recovery manager needs a mode manager that supplies to it the failure recovery strategy that is appropriate to current operational conditions. In accordance with the operational mode design pattern, the following abstract interface is defined to characterize mode managers for the failure recovery manager:

```
interface FrModeManager {
  RecoveryStrategy* getRecoveryStrategy();
  void loadRecoveryStrategy(RecoveryStrategy* r, int mode);
}
```

Assume now that the failure recovery manager needs a total of N operational modes. A concrete mode manager can then be defined as follows:

```
class ConcreteFrModeManager: public FrModeManager {

    RecoveryStrategy* implementer[N];
    int currentMode;
    . . .

    RecoveryStrategy* getRecoveryStrategy() {
       return implementer[currentMode];
    }
    . . .

    void loadRecoveryStrategy(RecoveryStrategy* r,
                                           int mode) {
       implementer[mode] = r;
    }
    . . .

    void updateMode() {
       . . .    // update currentMode
    }
}
```

The mode manager maintains a list of implementers and has at any given time a "current implementer" that is returned to the failure recovery manager. The current implementer is determined by method `updateMode` which contains the mode switching logic. The decision as to whom is responsible for calling `updateMode`, and therefore potentially triggering a mode switch, is of course application dependent. However, there are at least two typical mechanisms. In one case, it is desired to give the mode manager the chance to update its mode on a periodic basis (normally, once per AOCS cycle). This is best done by turning the mode manager into an active component. In a second typical situation, it is desired to link the mode switch to some other change within the application (perhaps, to a change of mode in another mode manager). This can be done by using the monitoring through change notification design pattern. The mode manager must then implement interface `PropertyMonitor` and its method `updateMode` is called by method `propertyChange`.

The failure recovery manager uses its mode manager as shown in the following pseudo-code:

```
class FailureRecoveryManager {

    FrModeManager* modeManager;
    . . .

    void activate() {
        RecoveryStrategy* r =
                        modeManager->getRecoveryStrategy();
        r->doRecovery();
    }
    . . .

    // Load mode manager plug-in component
    void setModeManager(FrModeManager* mm) {
        modeManager = mm; }
}
```

Note that the failure recovery manager sees the mode manager as a plug-in component that is loaded during application initialization. The advantage of having the mode manager characterized by an interface is that the failure recovery manager is thus insulated from the implementation of the mode manager.

The configuration code that is executed during application initialization for a simple case of a mode manager with two operational modes looks like this:

```
// Create the failure recovery mode manager
FrModeManager* mm = new ConcreteFrModeManager();

// Create recovery strategies
RecoveryStrategy* r1 = new ConcreteRecoveryStrategy1();
RecoveryStrategy* r2 = new ConcreteRecoveryStrategy2();

// Configure failure recovery mode manager
mm->loadRecoveryStrategy(r1,1);
mm->loadRecoveryStrategy(r2,2);
```

```
// Create and configure the failure recovery manager
FailureRecoveryManager* frm = new FailureRecoveryManager();
frm->setModeManager(mm);
```

It should be pointed out that both the failure recovery manager and the mode manager can be reconfigured dynamically during application normal operation. This would typically be done by telecommand.

This example has considered the failure recovery manager. This is a core component and its mode manager interface and default implementation corresponding to common mode switching logics would normally be predefined by the framework. However, the same steps can be taken by application developers to endow any application-specific component with mode-dependent behaviour.

Table 31. Implementation case example for the failure recovery mode manager

Objective	Create and configure a mode manager for the failure recovery manager.
Description	The failure recovery manager operates on a failure recovery strategy. Some applications require the recovery strategy to be different in different mission phases. This effect is modeled by making the recovery manager mode-dependent using the operational mode design pattern. This IC shows how the corresponding mode manager can be created and configured during the application initialization phase.
Framelet	Failure Recovery Framelet
	Operational Mode Framelet
Constructs	`FailureRecoveryManager` core component
	`RecoveryStrategy` abstract class
Hot-Spots	Mode Implementer Hot-Spot
	Mode Manager Hot-Spot
Related ICs	None
Implementation	- define abstract interface to characterize the mode manager.
	- create a concrete class implementing the interface.
	- load the concrete implementer in the mode manager.
	- load the mode manager into the mode-dependent component.

19.5 Implementation Case Example – 2

Consider the implementation case described in table 32. This section shows how it can be worked out during the framework design definition stage. Its objective is to construct a recovery action to reset a generic component. Framework components implement interface `Resettable` and therefore expose a method `reset` whereby they can be asked to reset their internal state (see section 10.2). This is exploited to construct a recovery action that has the component to be reset as a plug-in component. When the recovery action is executed – when its `doRecovery` method is called – the recovery action simply calls method `reset` on the plug-in component. This is illustrated in the UML diagram of figure 19.2.

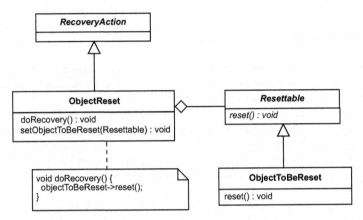

Fig. 19.2. Class diagram of a recovery action to reset a component

The reset recovery action is offered as a default component by the framework. Application developers that need to use it will have to write instantiation code like the following:

```
// Object to be reset
Resettable* obj = new ObjectToBeReset();

// Create and configure the recovery action
ObjectReset* objectReset = new ObjectReset();
objectReset->setObjectToBeReset(obj);
```

After being created and configured, the `objectReset` recovery action can be treated as an abstract recovery action and no other part of the AOCS application need know its concrete type or its implementation.

Table 32. Implementation case example for the failure recovery framelet

Objective	Construct a recovery action to reset a component.
Description	Framework components implement interface `Resettable` and are therefore capable of resetting their internal state. Resetting a component is a common response to a failure condition in the component. This IC shows how recovery action can be built to perform such a reset on a generic component.
Framelet	Failure Recovery Framelet
	System Management Framelet
Constructs	`RecoveryAction` abstract class
	`Resettable` interface
Hot-Spots	Recovery Action
Related ICs	None
Implementation	- Subclass `RecoveryAction` to override its `doRecovery` method to implement the reset action
	- Plug component to be reset into the newly created recovery action

19.6 Alternative Implementation

The failure detection design pattern described in the previous chapter foresees a single level of failure detection. This means that regardless of where they are found, failures are always treated in the same manner: they give rise to events that are stored in the same failure event repository. This in particular means that failures encountered by the failure recovery manager while processing the failure event repository are also stored in the failure event repository itself. Thus, the failure recovery manager may end up processing failures that it has itself generated as a result of its implementing a failure recovery strategy.

Clearly, this situation could potentially give rise to the same failure events being generated cyclically. Note, however, that whether or not this situation arises depends on which recovery actions are associated to which failures. Since recovery actions are plug-in components that the application developer can load when it configures the framework, their judicious choice can avoid the problem from arising. This is the approach assumed at present.

An alternative approach recognizes two levels of failure detection and recovery. To each level, a distinct failure event repository is associated. The framework normally runs in level 1. Failures detected at this level are stored in `FailureEventRepository_1`. When the failure recovery manager runs, the framework switches to level 2 and failure events are sent to `FailureEventRepository_2`. Thus, the danger of circularity is removed because while the failure recovery manager processes `FailureEventRepository_1`, any failures that are encountered are sent to `FailureEventRepository_2`.

The drawback of this alternative approach is that failures sent to `FailureEventRepository_2` remain unprocessed. One could envisage a level 2 recovery manager that is responsible for `FailureEventRepository_2`. From an implementation point of view, multiple failure recovery manager can be simply created by instantiating multiple instances from class `FailureRecoveryManager`. The main problem is that, if circularity is again to be avoided, a third level of failure detection is required. In short, if circularity is to be avoided by design, then it is inevitable that there will be some failure events that cannot be processed within the framework. The design and architectural solutions proposed by this and the failure detection framelets would allow this to be done but the recommended approach is to use a single level of failure detection and recovery and to avoid circularity by careful selection of the recovery actions associated to each failure.

19.7 Reusability

Reusability of failure recovery is achieved within the AOCS domain by decoupling the management of the recovery responses from their implementation. Two orthogonal levels of failure responses are provided. Recovery actions provide localized failure handling capabilities. Recovery strategies implement global failure recovery responses. Recovery strategies and recovery actions can be defined without affecting any part of the recovery handling code. In this way, the failure recovery manager becomes fully reusable across applications.

The failure recovery design pattern is at the core of this framelet. Its reusability outside the AOCS domain depends on whether the notions of recovery strategy and recovery action are adequate to capture the failure recovery philosophy of the domain. Since they are fairly general, this will often be the case. The implementation of the design pattern at architecture and code level depends, among other things, on the identification of the failure reports upon which recovery actions operate with failure events, and of the sets of failure reports upon which the failure strategies operate with the failure event repository. Reuse at architecture and code level therefore implies reuse of the event mechanisms proposed by the intercomponent communication framelet.

Table 33 summarizes the constructs exported by the framelet that have been defined or identified in this chapter.

Table 33. Overview of failure detection framelet

Design Patterns	
Failure Recovery Management	Separate the management of the failure recovery process from the implementation of the failure recovery responses.
Abstract Classes	
RecoveryAction	Base class for all classes encapsulating a local response to a single failure report (recovery action).
RecoveryStrategy	Base class for all classes encapsulating a global response to a set of failure reports (recovery strategy).
Core Components	
FailureRecoveryManager	Component to manage the execution of failure recovery strategies.
Default Components	
Recurring Recovery Actions	Components encapsulating commonly used recovery actions (object reset, system reset, mode change, reconfiguration, etc)
Recurring Recovery Strategies	Components encapsulating commonly used recovery strategies (system reset on too many failure reports, local recovery actions, etc)
Hot-Spots	
Recovery Strategy	Applications must define global responses to the sets of failure reports that can be generated during one failure detection cycle by subclassing RecoveryStrategy.
Recovery Action	Applications must define responses for each failure check by subclassing RecoveryAction.

20 The Telecommand Management Framelet

Telecommands encode actions to be performed on or by the AOCS software. Telecommands in traditional systems are represented by strings of bytes of which the first one is an identifier that defines the type of telecommand and that is followed by one or more bytes representing the data associated to the telecommand. Telecommand management is usually delegated to a dedicated task that, essentially, implements a case construct where, depending on the telecommand identifier, certain actions are taken. This solution is obviously not acceptable from the perspective of a framework which must explicitly recognize telecommands as hotspots and, for reasons of extensibility and reusability, should separate their management from their implementation. Separation is essential to achieving reusability since the implementation of the telecommands is necessarily application-dependent. The telecommand framelet proposes a design solution to achieve this separation.

20.1 The Telecommand Management Design Pattern

An application instantiated from the AOCS Framework is built as a collection of components that expose certain interfaces. The telecommands use these exposed interfaces to perform their assigned tasks. In practice, a telecommand is a component that exposes a basic execute method which performs the actions associated with the telecommand itself. The execute method is called by the telecommand manager which in this way remains completely insulated from any knowledge of what the telecommand does and how it does it.

Consider as an example the case of a telecommand that must trigger a self-test on a sensor. The sensor is implemented as an instance of class AocsUnit. As discussed in section 15.2, this class exposes a doSelfTest method. The telecommand's execute method then simply calls doSelfTest on the sensor component.

Execution is the basic operation that can be performed upon a telecommand but the way telecommands are typically used within the AOCS domain requires that they be endowed with the capability to perform other operations. Telecommands for instance may have a time tag specifying when they are to be executed. The base telecommand class should therefore declare a getTimeTag method to let the telecommand manager retrieve the time tag (which is part of the data associated to the telecommand). Often, there is also a requirement to enable and disable indivi-

A. Pasetti: Software Frameworks, LNCS 2231, pp. 243-252, 2002.
© Springer-Verlag Berlin Heidelberg 2002

dual telecommands and then methods like `setEnableStatus` and `isEn-abled` should also be declared by the telecommand class. Finally, telecommands are critical operations that should be executed only if the operational condition of the spacecraft warrants it. In order to ensure that this is the case, it is necessary that telecommands offer a `canExecute` method that verifies that the telecommand can indeed be executed at a certain point in time.

Given this conceptualization of telecommands, a telecommand management design pattern can be introduced to address the problem of separating telecommand management from telecommand implementation. It is closely based on the command design pattern and is illustrated in figure 20.1.

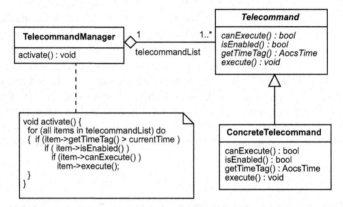

Fig. 20.1. Class diagram of the telecommand management design pattern

The telecommand manager maintains a list of pending telecommands and, when it is activated, it goes through the list and executes all telecommands in sequence. The telecommand pattern can also be seen as a refinement of the manager meta-pattern where telecommand execution is the pattern functionality, the telecommand manager is the functionality manager, and the `Telecommand` class decouples the functionality management from the functionality implementation. The telecommand manager then becomes one of the framework functionality managers.

The telecommand design pattern is instantiated within the framework and should normally not be visible to the framework users. There are at least three major issues that need to be considered in connection with its instantiation. Firstly, who should be responsible for activating the telecommand manager? As in the case of the other functionality managers, the most natural choice is to make the functionality manager into an active component in which case the component should implement interface `Runnable` and its `activate` method is then identified with the `run` method inherited from this interface.

Secondly, telecommand execution and loading are important events in an AOCS and there is usually a requirement that they be recorded in some kind of on-board log. The event mechanism of chapter 13 can serve this purpose. A telecommand event class should be created together with its associated event repository and the telecommand manager should be given the capability to create events after telecommands have been loaded and executed (see the implementation case example of section 13.5).

Thirdly, the telecommand concept is probably best implemented as an abstract base class rather than as an interface because some of its functionalities – for instance the management of the time tag and of the enable status – can be directly implemented at the level of base class. The telecommand action is coded in the implementation of method `execute` that is accordingly abstract and that constitutes a framework hot-spot (the *telecommand hot-spot*).

A possible pseudo-code for the `activate` method of the telecommand manager could be:

```
TelecommandList* tcList;    // list of pending telecommands

void activate() {
  for (all telecommands t in tcList) do
    if (t->getTimeTag()<currentTime)
    { if ( (t->isEnabled() )
         if ( (t->canExecute() )
         { t->execute()
                . . .    // create event to record TC execution
         }
         . . .        // unload telecommand
    }
}
```

With this implementation a telecommand that has come due for execution according to its time tag is unloaded regardless of whether it was executed or not (e.g. it is unloaded even if it was not executed because it was not enabled). Obviously, other solutions are possible but a point that is worth stressing is that the telecommand manager is a core component and therefore the choice about what to do with telecommands that are due for execution but are not enabled or are otherwise not ready to execute is a domain-level choice. This means that it should not be treated as a hot-spot where applications have the flexibility to choose an application-specific implementation. A choice should be made based on the prevailing custom in the AOCS domain but this is then hard-coded into the telecommand manager component and has to be accepted by all applications that use the framework (or at least by all applications that use this framelet). In terms of the functionality concept of section 7.2, the logic for the telecommand manager would be described by do-functionalities.

As was already noted in section 17.4, there is an obvious similarity with the procedure to execute manoeuvres which, in an alternative design solution, could be exploited to merge the two concepts of telecommand and manoeuvre.

This solution achieves a high degree of reusability of both individual telecommands and of the telecommand management software. This is a consequence of the complete de-coupling between the content of telecommands and the logic required to execute them. Individual telecommands can be re-used because they consist of calls to public methods – whose syntax presumably does not change from mission to mission. The telecommand manager being independent of telecommand content can also be re-used without changes. The independence of the telecommand manager from the content of telecommands also means that new telecommands can be added without any impact on the telecommand management software thus achieving the objective of ease of extensibility.

20.2 The Telecommand Transaction Design Pattern

In current systems, execution failure for a telecommand is reported in telemetry and it is then left to the ground to take whatever corrective action is appropriate. The AOCS Framework introduces the *telecommand transaction design pattern* to address the problem of treating some telecommands or some sequences of telecommands as *transactions* where the term "transaction" is used in the same sense in which it is used in database systems to designates an atomic operation that can either succeed or fail and that, in case of failure, restores the initial state of the system. Thus, transactions are safe because even in case of failure they leave the AOCS software in a consistent state.

Consider for instance a situation where the AOCS is in mode A and must make a transition to mode B. Suppose also that mode B requires units 1, 2 and 3 to be switched on. One would use the following sequence of telecommands to perform the desired transition:

- switch on unit 1;
- switch on unit 2;
- switch on unit 3;
- switch to operational mode B.

If, say, the second telecommand fails, two corrective actions should be performed:

- abort the telecommand sequence, and
- switch off unit 1.

Note that simply aborting the telecommand sequence would leave the AOCS in an inconsistent state where unit 1 (which should be switched off in mode A) remains powered. Consistency in case of intermediate failures will be assured by treating the telecommand sequence as a transaction.

The telecommand transaction design pattern calls for the introduction of a class `TransactionTelecommand` that extends calls `Telecommand` with an `unExecute` method that "undoes" the actions of the telecommand. The resulting class structure is shown in figure 20.2 where, for simplicity's sake, only the `execute` method is shown in the `Telecommand` class.

Note that, as shown in the class diagram, telecommand transactions must be linked in a chain to allow their manager to recursively undo all telecommands in the same transaction. In instantiating the pattern, the main problem is to ensure that the telecommand manager can treat all telecommands in the same way regardless of whether they are transaction or individual telecommands. This probably means that the logic for checking the execution status of a telecommand and, if necessary, for unexecuting a failed telecommand should be placed in the implementation of methods execute and unexecute of class `TelecommandTransaction`.

Concrete transaction telecommands must provide implementations for both `execute` and `unExecute`. In some cases, unexecution is either not possible or not desired. In this case, the telecommand transaction becomes merely a *telecommand sequence*, namely a sequence of telecommands that are executed as a single telecommand but for which there is no guarantee of system integrity in case of partial or complete failure.

The telecommand transaction design pattern introduces the possibility of recursion since a call to method `execute` can now be recursive. The maximum depth of the recursion is however easy to bound and is given by the maximum number of telecommands that are chained into a single telecommand transaction.

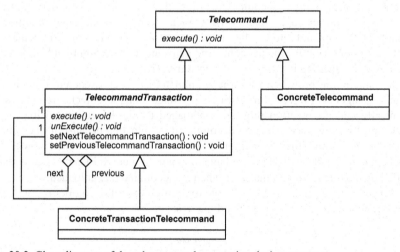

Fig. 20.2. Class diagram of the telecommand transaction design pattern

20.3 Telecommand Loading

By definition, telecommands must be loaded dynamically into the AOCS computer memory. The loading mechanism is application-dependent. In one common mechanism the telecommand is loaded via DMA under the control of hardware that is external to the AOCS computer. The completion of a load operation is indicated by an interrupt to the AOCS software. In a second common loading mechanism the telecommand words or bytes are loaded by the AOCS software via I/O commands. The arrival of a new word or byte is indicated by an interrupt to the software.

Since telecommand loading is application-specific, the framework cannot provide a generic telecommand loader component. Telecommand loading then represents a framework hot-spot (the *telecommand loader hot-spot)* and some adaptation mechanism should be devised to model it. The simplest solution is to associate an abstract interface that concrete telecommand loaders then have to implement. The basic function of a telecommand loader is to process the raw telecommand as described by the string of bytes that are uplinked by the ground station, use them to construct an object of type `telecommand`, and then load the resulting telecommand object into the telecommand manager who will be responsible for executing it.

The telecommand loader therefore interacts, on the one side, with the low-level mechanism for receiving the raw telecommands and, on the other side, with the telecommand manager. The interaction with the former is best modelled as a response to an interrupt, probably mediated by the operating system. This turns the telecommand loader into an active component, namely a component whose activation is controlled by a scheduler external to the framework. The interaction with the telecommand manager arises when the telecommand loader loads a newly assembled telecommand into it. This simply means that the telecommand manager is a plug-in component for the telecommand manager. The telecommand loader must then expose a method like `setTelecommandManager`.

At some point, probably after a telecommand has been executed, the telecommand manager will have to unload it. Since telecommand objects are created dynamically and the creation process is managed by the telecommand loader, the latter must be informed when a telecommand is unloaded because it needs to release the resources (mainly memory) allocated to the telecommand object. The telecommand loader will therefore have to expose a method like `release` that is called by the telecommand manager when a telecommand is unloaded.

After this long premise, it is possible to propose the following definition for the interface characterizing telecommand loaders:

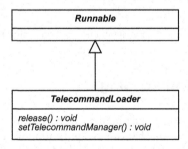

In summary, a complete telecommand processing cycle might proceed as follows:

- an interrupt signals the arrival of a new telecommand (there may be a single interrupt for the whole telecommand or an interrupt for each telecommand word)
- the telecommand loader reacts to the interrupt by constructing an object of the appropriate telecommand class
- the telecommand loader loads the newly assembled telecommand into the telecommand manager
- eventually, the telecommand manager executes the telecommand or perhaps discards it if it was not enabled or if operational conditions did not allow its execution
- the telecommand manager informs the telecommand loader that the telecommand has been discarded (by calling its `release` method)
- the telecommand loader releases any resources that had been allocated to the telecommand object

20.3.1 Implementation Considerations

The first point to mention relates to the implementation of concrete telecommand loaders. The memory allocation process for the telecommand objects probably cannot be managed using predefined `new` operators because these normally have an unpredictable behaviour. In view of the hard real-time nature of AOCS applications, the telecommand loader will have to define its own memory allocation mechanism in a manner that ensures timing predictability.

A second point concerns the content of a telecommand load from the ground to the satellite. In the concept proposed here, a telecommand is an object made up of data and code. The link between the data and the code is provided by a virtual function table. Two types of telecommand loading can be anticipated:

- *Data Load*: The code for the telecommand is already present in the AOCS software and only the data part of the telecommand object is physically loaded into memory.
- *Full Load*: Both the code and the data for the telecommand object are loaded.

The two telecommand load types are illustrated in figure 20.3 for the case of a telecommand class that defines three methods. In a data load, only the lightly shaded part of the figure is physically loaded onto the AOCS computer memory. In the full load case, both the light gray and dark gray parts are loaded. The example in the figure assumes that the code for all methods required by the telecommand must be loaded. In practice, some methods will be able to rely on code already present in the AOCS computer (typically, this is because such methods are inherited without changes from a superclass of the concrete telecommand class that is already present in memory).

Normally, an AOCS application will include some predefined telecommand classes implementing common telecommand actions. For such telecommands, only the data need to be loaded. Unusual or unforeseen telecommands need to be

loaded in full. Typical telecommands that are used in most AOCS applications and
that must be provided as default components by the framework include:

– change the operational mode of a component
– reset the AOCS software
– command a reconfiguration to a reconfiguration manager
– patch a memory area

In the traditional telecommand concept, an identifier at the head of the telecommand defines the action associated to the telecommand. In the concept proposed here, the pointer to the virtual function table plays the role of the identifier.
The main advantage of this concept is its higher expressiveness. Telecommands
can be associated to virtually any class which is present on board or they can also
be defined anew without any impact on the telecommand management logic. The
price paid for this extra flexibility is bigger telecommand size. The classical telecommand identifier only takes one byte whereas the virtual table pointer takes
four bytes. Additionally, if the telecommand class is not already present on board,
code must be uploaded as well as data. This however will occur rarely (frequent
telecommands should have their corresponding classes predefined on board) and
the complexity of the code should not be overstated since in most cases it will
simply consist of method calls to other existing classes.

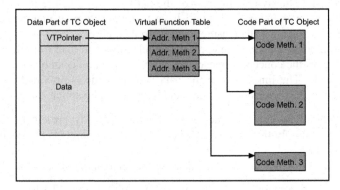

Fig. 20.3. Data elements making up a telecommand

20.4 Implementation Case Example

Consider the implementation case described in table 34. This section shows how it
can be worked out during the framework design definition stage. Its objective is to
create a telecommand to reconfigure a redundant unit. According to the design
pattern of section 16.2, reconfigurations are always handled through reconfiguration managers, i.e. components that implement interface Reconfigurable.
Entities that perform the reconfiguration (as opposed to entities that use the servi-

ces of the reconfigurable components) always see the object of the reconfiguration as an instance of type `Reconfigurable`.

The telecommand design pattern prescribes that to every concrete telecommand, there must correspond a class derived from base class `Telecommand`. The telecommand class for this implementation case can be defined as follows:

```
class ReconfigTc : public Telecommand {
   Reconfigurable* r;
   . . .
   void execute() {
      r->reconfigure();
   }
}
```

Note that the sensor to be reconfigured is seen by the telecommand simply as an instance of type `Reconfigurable` and therefore this telecommand can be used to reconfigure any redundant functionality, not just redundant sensors.

Method `execute` does not perform any check on the success of the reconfiguration. This could be easily added if desired.

Reconfigurations are common operations in the AOCS domain and therefore it is likely that class `ReconfigTc` would be linked into the AOCS executable. The reconfiguration telecommand therefore would simply consist of the address of the reconfiguration manager responsible for the redundant sensors. In the terminology of the previous section, it would be constituted only of a data load.

Table 34. Implementation case example for telecommand framelet

Objective	Construct a telecommand for reconfiguring a redundant sensor.
Description	Redundant sensors can be reconfigured dynamically by ground command. This IC shows how a telecommand can be constructed to perform such a reconfiguration.
Framelets	Telecommand Framelet
	Reconfiguration Framelet
Constructs	`Telecommand` abstract base class
	`Reconfigurable` interface
Hot-Spots	Telecommand hot-spot
Related ICs	None
Implementation	- Represent the redundant sensor as an istance of type `Reconfigurable`
	- Subclass `Telecommand` to override the `execute` method to call method `reconfigure` on the reconfigurable component

20.5 Reusability

Reusability of the telecommand concept within the AOCS domain is guaranteed by the separation between the telecommand content and its management.

All embedded control systems need to be able to process external commands coming either from a human operator or from some higher level supervisory computer. The problem addressed by the framelet therefore arises more widely in the entire embedded control domain. The solution it proposes seems to be generic enough to be suitable beyond the AOCS domain. This is probably true at both design and architectural level. At code level, the internal logic of the telecommand manager may be AOCS-specific as are the default components implementing particular telecommands and telecommand loaders.

Table 35 summarizes the constructs exported by the framelet that have been identified or defined in this chapter.

Table 35. Overview of telecommand framelet

Design Patterns	
Telecommand Management	Separate management of TCs from their implementation - refines manager meta-pattern and command pattern
Telecommand Transaction	Handle sequences of TCs as single transaction that succeed or fail as a single entity.
Abstract Classes	
Telecommand	Encapsulate telecommand abstraction
TelecommandTransaction	Encapsulate telecommand transaction abstraction
TelecommandLoader	Encapsulate telecommand loader abstraction
Core Components	
TelecommandManager	Implement logic to manage TC execution
Default Components	
Concrete Telecommands	Commonly used telecommand classes: mode change, reconfiguration, system reset, memory patch, etc
Concrete Telecommand Loaders	Implementations of commonly used telecommand loading mechanisms: DMA loading, word-by-word loading
Hot-Spots	
Telecommand	Applications must define their own telecommands
Telecommand Loader	Applications must define their own telecommand loading mechanism

21 The Telemetry Management Framelet

Telemetry data are generated cyclically by the AOCS for transmission to the ground station. They represent a subset of the software state of the AOCS and are used on ground to check AOCS operation. Telemetry data are transmitted in frames where a telemetry frame represents the set of data that are sent by the AOCS to the ground in one telemetry transmission cycle. In existing AOCS applications, telemetry processing is usually performed by a dedicated task centered around a component that collects the telemetry data from other components in the software and formats them in packets suitable to be forwarded to the ground station. This solution is obviously not reusable because the type and format of telemetry data vary from mission to mission.

The design solution offered by this framelet assumes that the AOCS software is organized as a collection of components with each component potentially capable of writing its own state to the telemetry stream. A telemetry manager component is responsible for managing the flow of telemetry data and is one of the framework functionality managers. The structure and content of the telemetry data becomes a framework hot-spot.

21.1 The Telemetry Management Design Pattern

This design pattern is introduced to address the problem of separating the management of telemetry data from the layout and content of the telemetry frames. It is based on the manager meta-pattern and is illustrated in the class diagram of figure 21.1.

The telemetry manager maintains a list of references to components of type `Telemeterable`. Telemeterable components are components that are capable of writing their own internal state to the telemetry stream. The telemetry manager additionally maintains a reference to the *telemetry stream*. The telemetry stream is the data sink to which telemetry data are written. It represents the channel through which telemetry data are forwarded to the ground station.

The characteristics of concrete telemetry streams vary widely across AOCS applications and therefore no generic telemetry stream component can be provided. The AOCS Framework characterizes telemetry streams through the abstract interface `TelemetryStream`. From the point of view of its clients, a telemetry stream is essentially a component to which individual data items can be written. Its interface can therefore be defined as follows:

A. Pasetti: Software Frameworks, LNCS 2231, pp. 253-262, 2002.

TelemetryStream

write(flloat) : void
write(double) : void
write(int) : void
write(short) : void
write(bool) : void
streamManagement() : void

The telemetry stream should offer write methods for every primitive type. These write methods are then used by the telemeterable components to write the value of their internal variables to the telemetry stream. Additionally, it should offer methods for performing stream management operations such as resetting the stream, flushing it, etc.

Telemetry data are often collected in DMA mode by dedicated hardware. A common type of telemetry stream therefore consists of a component that writes the telemetry data to a DMA buffer. Such a component could be provided as a default component by the framework.

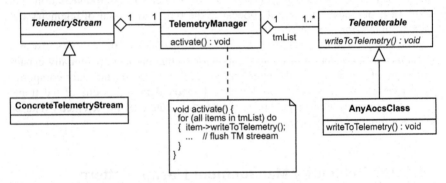

Fig. 21.1. Class diagram of the telemetry management design pattern

In considering the instantiation of this pattern in the framework, the first issue to be addressed is the activation of the telemetry manager. Since telemetry data are typically collected cyclically and in synch with the basic AOCS control cycle, the simplest solution is to make the telemetry manager an active component implementing interface `Runnable` and to identify its `activate` method with method `run`.

The telemeterable interface needs to declare rather more methods than just `writeToTelemetry`. For instance, it is necessary for the telemetry manager to know what the size of the telemetry image generated by a given component is so as to verify that it is compatible with the telemetry channel capacity. This could be done by adding to the telemeterable interface a `getTelemetryImageLength` method.

Usually, the format and content of the telemetry data depends on operational conditions. There are some mission phases when only basic telemetry information is required by the ground and there are others where instead more detailed information is needed. This variability can be handled within the framework by making the telemetry manager a mode-dependent component using the design pattern of chapter 12. In this case, the strategy associated to an operational mode of the telemetry manager is the list of telemeterable components. The telemetry mode manager then is a component that exposes a method called `getTelemetryList` and that can return the telemetry list appropriate to current operational conditions. The telemetry stream is usually fixed and therefore there is no need to make its selection mode-dependent. If this were the case, then the telemetry mode manager should have two strategies: the telemetry list and the telemetry stream and it should expose a second method called `getTelemetryStream` that returns the telemetry stream appropriate to current operational conditions.

One final implementation issue to be considered concerns the actual writing of the telemetry data to the telemetry stream. Should it be done by the telemetry manager or by the telemeterable components? The second solution is preferable because it minimizes the coupling between the telemetry manager which should be application-independent and the application-dependent components as represented by the concrete telemetry streams and the concrete telemeterable component. The telemetry stream then becomes a parameter that is passed to a telemeterable component in the call to method `writeToTelemetry`. The telemeterable component uses the write methods exposed by the telemetry stream to send its internal state to the telemetry stream.

A sample implementation of the activation method of the telemetry manager becomes:

```
TelemetryModeManager* tmModeManager;
TelemetryList* tmList;
TelemetryStream* tmStream;

void activate() {

    // Query TM mode manager for the TM list
    tmList = tmModeManager->getTelemetryList();

    for (all telemeterable items t in tmList) do
    {   size = t->getTelemetryImageLength();
        if (TM image size fits in TM stream capacity)
            t->writeToTelemetry(tmStream);
        else
            . . .   // handle error
    }
}
```

This design pattern covers two hot-spots. The definition of the layout and content of the telemetry image of telemetry components – through the implementation of interface `Telemeterable` – is application-specific and represents the first hot-spot (the *telemetry hot-spot*). The definition of the concrete telemetry stream – through the implementation of interface `TelemetryStream` – is also application-specific and represents the second hot-spot (the *telemetry stream hot-spot*).

Finally, it may be mentioned that there is some similarity between the telemetry design pattern and the message-logging pattern of reference [54]. The main difference lies in the fact that the latter gives components control over *when* data logging should be performed whereas the telemetry pattern creates an external component – the telemetry manager – that is responsible for triggering the forwarding of telemetry data.

21.2 Implementation Case Example

Consider the implementation case described in table 36. This section shows how it can be worked out during the framework design definition stage. Its objective is to make it possible to include data from a given component in telemetry. Let this component be an instance of the following class:

```
class aClass {
    type1 var1;
    type2 var2;
    . . .
}
```

It is assumed that the data the component must contribute to the telemetry stream are the values of var1 and var2. Application of the telemetry design patterns dictates that the component be made to implement interface Telemeterable and that its writeToTelemetry method be implemented so as to write the values of var1 and var2 to the telemetry stream. The class definition of the selected component must then be modified as follows:

```
class aClass : Telemeterable {
    type1 var1;
    type2 var2;
    . . .
    void writeToTelemetry(TelemetryStream* tmStream) {
        tmStream->write(var1);
        tmStream->write(var2);
    }

    int getTelemetryImageLength () {
        return sizeof(type1)+sizeof(type2);
    }
}
```

The implementation of method writeToTelemetry assumes that interface tmStream defines operations that are suitable to write variables of type type1 and type2 to the telemetry stream. In practice, it may be necessary to perform some type conversion on the variables to be written to the telemetry stream.

The modification to class `aClass` simply shows how our component can be endowed with the ability to write itself to the telemetry stream. Actual inclusion of its state in the telemetry data requires that the component be included in a telemetry list that is then loaded into the telemetry manager. This means that during the framework instantiation process, the following code must be present:

```
TelemetryList* tmList;
aClass* aComponent;
TelemetryManager* tmManager;
. . .
tmList->addItem(aComponent);
. . .
tmManager->loadTmList(tmList);
```

The pseudo-code assumes that a dedicated type `TelemetryList` is available to represent a list of telemeterable components. A telemetry list must be instantiated and the component must be loaded into it. The telemetry list must then in turn be loaded into the telemetry manager which will then take care of triggering the writing of the component's state to the telemetry stream.

The instantiation code shown above assumes that the telemetry list is directly loaded into the telemetry manager. In fact, the telemetry manager is a mode-dependent component which receives its telemetry list from a telemetry mode manager. In a more realistic example, the telemetry list would be loaded into the telemetry mode manager and the telemetry mode manager would then be loaded into the telemetry manager.

Table 36. Implementation case example for telemetry framelet

Objective	Include information from a given component in telemetry.
Description	Telemetry data are used to report the state of the AOCS software. This IC shows how data from a given component can be included in the telemetry stream.
Framelet	Telemetry Framelet
Constructs	`Telemeterable` interface
	`TelemetryStream` interface,
	`TelemetryList` component
Hot-Spots	Telemeterable hot-spot
Related ICs	None
Implementation	- Make the component implement the `Telemeterable` interface
	- Implement `writeToTelemetry` to send the desired data to the telemetry stream.

21.3 Functionality List Example

Chapter 7 introduced the functionality concept as a way of describing the framework *a posteriori* and from the point of view of its prospective users. Table 37 shows an incomplete list of the functionalities that originate in the telemetry framelet. Functionalities are intended to be written after the framework development has been completed. This book only describes the framework at concept level and therefore some of the functionalities in the table are not reducible to the constructs presented in this chapter.

It should also be noted that functionalities are defined at framework level, not at framelet level. This is because they describe the framework as a fully implemented artifact and framelets have well-defined boundaries only at design level. It is therefore not always possible to allocate functionalities to framelets. More precisely, functionalities arise in a certain framelet but they may enter into relationships with functionalities originating in other framelets or may be mapped to architectural constructs that originate in other framelets.

The table lists the functionalities together with their relationships to other functionalities and mappings to architectural constructs. Where appropriate, below each functionality, a comment box is presented that discusses its content.

Table 37. Functionality list from the telemetry framelet

DF1	The framework provides a generic and customizable component – the telemetry manager component – to control the acquisition of telemetry data and their forwarding to the telemetry stream.
	***expands-to** DF1.1 to DF1.6*
Comment:	This is a high-level do-functionality. It is not mapped to any architectural constructs because it is expanded to other, lower-level, functionalities.
DF1.1	The telemetry manager is an active component.
	***is-implemented-by** TelemetryManager component*
DF1.2	The telemetry manager provides reset and configuration services.
	***is-implemented-by** TelemetryManager component* ***uses** CF1 from the system management framelet (reset services)* ***uses** CF2 from the system management framelet (configuration services)*
Comment:	This functionality refers to the fact that the telemetry manager component in its final form implements interfaces Resettable and Configurable exported by the system management framelet and therefore uses the reset and configuration services defined by this framelet (these are referred to as can-functionalities CF1 and CF2).
DF1.3	The telemetry manager is a telemeterable component
	***is-implemented-by** TelemetryManager component* ***uses** CF3 from this framelet*
Comment:	This functionality refers to the fact that the telemetry manager component in its final form implements interface Telemeterable
DF1.4	The telemetry manager maintains a set of telemetry lists. A telemetry list is a list of components whose state must be included in telemetry. At time, only one telemetry list is active. When it is activated, the telemetry manager directs the components in the active list to send their state to the telemetry stream.
	***is-implemented-by** TelemetryManager component*

Table 37. (cont.)

DF1.5	The telemetry manager checks that the size of the telemetry image associated to the current telemetry list is compatible with the capacity of the telemetry stream. If this is not the case, then a failure event is generated. ***is-implemented-by*** `TelemetryManager` *component* ***uses** DF1 from intercomponent communication framelet (error reporting)*
Comment:	The use relationship indicates that the generation of the failure event is made using a the standard event generation mechanism of the framework
DF1.6	The telemetry manager sends the telemetry data to a single telemetry stream. ***is-implemented-by*** `TelemetryManager` *component*
Comment:	This is the kind of functionality that might give rise to an is-incompatible-with mapping with an application requirements: applications that send telemetry data to more than one telemetry stream, are outside the framework domain.
CF3	Any component can be made telemeterable, namely it can become a component whose state can be sent to telemetry. ***matches** the Telemeterable hot spot*
CF4	The telemetry manager can be customized to send any combination of component states to the telemetry stream. ***expands-to** CF4.1 and CF4.2*
CF4.1	The telemetry manager can manage any number of telemetry lists. ***matches** the Telemeterable List Hot-Spot*
Comment:	The telemeterable list hot-spot was not presented here because it arises during the architectural design phase when data structures are defined to hold the telemeterable components to be sent to the telemetry stream in one given frame.
CF4.2	The telemetry manager can be customized to use any algorithm to select the active telemetry list ***is-implemented-by** OF4.1* ***matches** the Telemetry Mode Manager Plug-in Hot-Spot* ***uses** DF1 from operational mode framelet (mode management pattern)*
Comment:	The telemetry mode manager hot-spot was not presented in this chapter because it arises only during the architectural design phase. It was hinted at in the discussion in the chapter when it is mentioned that a requirement to have different telemetry formats in different mission phases can be satisfied by making the telemetry manager a mode dependent component. This can be done using the mode management design pattern. For this reason, the functionality is listed as using the corresponding functionality from the mode management framelet. Note that this is an example of a functionality that is underlain not by a concrete construct but by a design pattern.
OF4.1	The telemetry management framelet offers a telemetry mode manager component implementing a cycling mode management mechanism to cycle through a fixed number of telemetry lists. ***matches** the Telemetry Mode Manager Plug-In Hot-Spot* ***uses** OF1 from operational mode framelet (`CyclingModeManager`)* ***is-implemented-by** the* `CyclingTelemetryModeManager` *component*

Table 37. (cont.)

Comment:	Component `CyclingTelemetryModeManager` is not mentioned in the chapter because it is defined only in the architectural definition phase. It encapsulates a default mode switching logic for the telemetry mode manager. It is in turn derived from component `CyclingModeManager` exported by the operational mode framelet.
CF4	The telemetry manager component can be customized to send the telemetry data to any telemetry stream. *is-implemented-by* OF4.1 *matches* the Telemetry Stream Plug-in Hot-Spot
Comment:	This functionality refers to the fact that in the final implementation of the framework, the telemetry stream is a plug-in component for the telemetry manager. This gives rise to a dedicated hot-spot.
OF4.1	The telemetry management framelet offers a component encapsulating a DMA-based telemetry stream. This telemetry stream assumes that telemetry data are collected by a DMA mechanism from a pre-defined buffer. *matches* the Telemetry Stream Plug-in Hot-Spot *is-implemented-by* the DmaTelemetryStream component
Comment:	Component `DmaTelemetryStream` is a default implementation of interface `TelemetryStream`.
CF5	The telemetry manager can be customized to associate any recovery action to the failure check of DF1.5. *matches* the Recovery Action Plug-In Hot Spot
Comment:	As discussed in section 18.1, to each failure check a recovery action must be associated. Since the telemetry manager performs one such failure check (as described in functionality DF1.5), it must expose a hot-spot where the corresponding recovery action can be loaded. This hot-spot is only defined in the architectural definition phase.
OF6	The telemetry management framelet offers a component to allow a range of contiguous memory locations to be written to the telemetry stream. *matches* the Telemeterable Hot-Spot *is-implemented-by* the MemorySection component
Comment:	AOCS applications often have a requirement that it should be possible to send to telemetry an image of a memory segment. Component `MemorySection` is predefined by the framework to facilitate implementation of this requirement.

21.4 Alternative Implementation

A very elegant way to implement the proposed telemetry concept is to treat telemetry reporting as a form of *serialization* (in the Java sense of the term). Telemetry then becomes an output stream to which the state of components marked for inclusion in telemetry are serialized. This approach presents the following advantages:

- *Simplicity of Implementation*: As in Java, serialization can be handled either by default methods or by class-specific methods. If the default serializer is used, then serialization is effectively transparent to individual classes thus simplifying their development.
- *Strong Heritage*: Serialization is implemented in a number of languages and is therefore a well-understood process for which good architectural and implementation models exist. This would result in a robust and reliable implementation.
- *Symmetry between On-Ground and On-Board Telemetry Processing*: Serialization allows components to be fully reconstructed on ground. This achieves the objective of symmetric treatment between on-board and on-ground telemetry. Telemetry becomes a means of "cloning" AOCS components on ground with considerable advantages in terms of reliability and simplicity of ground control software.

The drawbacks of this approach are:

- *Need for Language Support*: An efficient implementation of serialization requires dedicated language support. Firstly, the language must make a distinction between data members that are part of the *persistent state* of a class, and should therefore be serialized, and *transient* data members that can be ignored during the serialization process. Secondly, meta-language support is required to let the serializer extract information about the number and type of the data members present in a class.
- *High Bandwidth Requirements*: Standard implementations of serialization tend to generate large amounts of data. Serialization introduces overheads because it does not packetize data in the most efficient manner (eg a boolean variable will take one or more bytes rather than just one bit) and because it adds to the stream possibly redundant information about class names. This requires a higher capacity for the telemetry channel which in turn translates into higher requirements on the bandwidth of the telemetry downlink.

The first drawback can be overcome depending on the language that is selected for the AOCS software. In the case of C++, for instance, standard libraries exist that provide language support for serialization. The second drawback is more serious because bandwidth is usually severely limited on spacecraft. This is the main reason why serialization was discarded as an implementation option for the telemetry design pattern.

21.5 Reusability

Reusability of the proposed telemetry concept within the AOCS domain is achieved by decoupling the management of the telemetry data collection from the format and content of the telemetry frames. The former is encapsulated in a telemetry manager component that is application-independent. Applications must provide concrete implementations of the `Telemeterable` interface.

Although the terminology used by the framelet is AOCS-specific, the problem it addresses and the solution it uses are not. The framelet could be ported to any other embedded system where there is a need for sending housekeeping data to an outside entity on a regular basis. Reuse can certainly occur at design level because the framelet design pattern is quite general. It could occur at architectural level as well because the definition of the external interfaces of the framelet components is not AOCS-specific. Some of the components – in particular the telemetry manager – can probably be reused as they are at code level.

Table 38 summarizes the constructs exported by the framelet that have been identified or defined in this chapter.

Table 38. Overview of telemetry framelet

Design Patterns	
Telemetry Management	Separate management of telemetry data from the format and content of telemetry frames.
Abstract Classes	
`Telemeterable`	Characterize a component whose state can potentially be included in the telemetry stream.
`TelemetryStream`	Encapsulate a generic telemetry stream.
Core Components	
`TelemetryManager`	Implement logic to manage collection of TM data and their dispatching to the ground.
Default Components	
Concrete telemetry streams	Telemetry stream writing telemetry data to a DMA buffer
Hot-Spots	
Telemetry	Applications must define the content and format of telemetry data. They do so by providing implementations of interface `telemeterable`.
Telemetry Stream	Applications must define their own telemetry stream. Thy do so by providing implementations of interface `TelemetryStream`.

22 The Controller Management Framelet

An AOCS will typically contain several control loops serving such diverse purposes as stabilizing the attitude of the satellite, stabilizing its orbital position, controlling the execution of attitude slews, or managing the satellite internal angular momentum. The objects implementing these control loops in traditional AOCS applications tend to implement the same flow of activities starting with the acquisition and filtering of measurements from sensors, continuing with the computation of command actions, and ending with their application to actuators designed to counteract any deviation of the variable under control from its desired value. Despite the ubiquity of closed-loop controllers and the similarities in their structure, existing AOCS applications have not developed any abstraction to represent them. Controllers are normally implemented by one-of-a-kind objects that hardcode the control algorithm and their management is not recognized as an explicit and separate activity.

This situation is unsatisfactory in a framework perspective where controllers must be seen as a major hot-spot and where there is corresponding need to decouple their management from their implementation and to allow easy insertion of the specific control algorithms required by a particular AOCS application. This framelet proposes a way of achieving this effects by introducing an abstract interface to characterize controller components and a functionality manager to control them. The two are linked together by a dedicated design pattern.

22.1 The Controller Design Pattern

This design pattern is introduced to address the problem of separating the management of closed-loop controllers from their implementation. It is based on the manager meta-pattern and is illustrated in the class diagram of figure 22.1.

The abstract interface `Controllable` represents a generic closed-loop controller. Its key method is `doControl` that directs the controller to acquire the sensor measurements, derive discrepancies with the current set-point, and compute and apply the commands for the actuators. Since closed-loop controllers can become unstable, a second key method `isStable` which is provided to ask a controller to check its own stability. The controller manager component is responsible for maintaining a list of components of type `Controllable` and for asking them to check their stability and, if the stability is confirmed, to perform their allotted control action.

A. Pasetti: Software Frameworks, LNCS 2231, pp. 263-269, 2002.

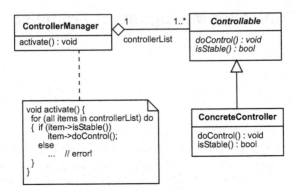

Fig. 22.1. Class diagram of the controller management design pattern

The controller design pattern is instantiated within the framework and is normally invisible to framework users. As usual when considering the instantiation of this type of pattern, the first issue to arise concerns the activation of the controller manager. In this case, as in the other cases of functionality managers in the AOCS Framework, the solution proposed by the framework at architectural level is to make the controller manager an active component and identify its `activate` method with the run method of interface `Runnable`.

In most cases, the control algorithms that are applied by each control loop depend on operational conditions. This is taken into account in the spirit of the framework by making the controller manager mode-dependent in the sense of chapter 12. A controller mode manager must then be built that manages a single strategy represented by the list of controller components that are associate to a given mode.

The implementation of interface `Controllable` gives rise to a major framework hot-spot (the *controllable hot-spot*) where application developers must define the structure and algorithms of their closed-loop controllers.

22.2 The Controller Abstraction

Closed-loop controllers can have many different structures but those that are used in the AOCS domain share some commonalities. These commonalities can be exploited to create a reusable and configurable component to encapsulate a generic closed-loop controller for the AOCS.

The model behind this component is shown in figure 22.2. The figure shows a classical closed-loop control system. The arrows represent signal flows and each arrow can represent several parallel flows. The boxes represent multi-input-multi-output transfer functions. The primary input to the controller is the measurement of the variable to be controlled. The measurement is compared with a reference value representing the desired value for the controlled variable. The difference between the two is the control error. This controller error is fed to a compensator –

a digital filter – that processes it and produces an output that may be corrected with a so-called feedforward signal. The result is a command for an actuator that in principle must cause the controlled variable to move closer to its target value. All input signals may be filtered before being processed.

More specifically, the inputs to a controller are:

- The *reference signals* that must be tracked by the controller
- The *measurements* from a sensor
- The *feedforward compensation signals* that are added to the compensator output and typically model known disturbances

The sensor measurement inputs are mandatory. The reference signals and feedforward compensation inputs are optional. If they are absent, the controller assumes that they are equal to zero. The controller outputs are:

- The *actuator commands* that represent the commands for the actuator
- The *compensator outputs* that represent the inputs to the compensator block
- The *control errors* that are obtained as the difference between filtered measurements and reference signals

The actuator command outputs are mandatory. The other two outputs are optional. Finally, a controller includes four transfer function blocks:

- The *compensator* which implements the control law
- The *measurement filter* that filters the measurement signals
- The *feedforward filter* that filters the feedforward signals
- The *reference signal filter* that filters the reference signals

Explicit specification of the compensator transfer function is mandatory. The other transfer function blocks need not be specified by the user in which case they are implicitly assumed to be equal to 1. In practice, the measurement filter block often represents an estimator.

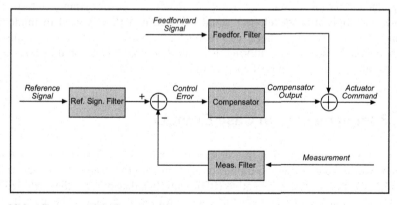

Fig. 22.2. Abstract model of a controller

The framelet offers a default component to encapsulate a controller of the form shown in the figure. Its class structure is shown in figure 22.3.

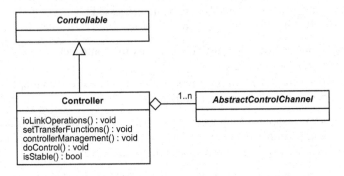

Fig. 22.3: Controller component implementing model of figure 22.2

This class exposes two major categories of methods. There are in the first place methods to link the controller inputs and outputs to user-defined locations in the data pools (`ioLinkOperations`). This is in accordance with the model for data exchanges among framework components of chapter 13. There are then methods to set the transfer functions in the controller (`setTransferFunctions`). Transfer functions are best represented as control channels and therefore these methods take as parameters variables of type `AbstractControlChannel` (see section 14.2).

Controllers implement interface `Controllable` and must therefore provide implementations of its fundamental operations `doControl` and `isStable`. A call to the former will result in all the control channels in the controller being asked to propagate their inputs and update their outputs. Controllers of the kind shown in figure 22.2 do not have any generic stability check associated to it. The implementation of method `isStable` therefore should simply return true. However, application developers could override it if they wished to implement some particular stability check. Alternatively, an indirect check on the controller stability can be done by monitoring the controller errors. This could be done using the monitoring mechanisms of chapter 11.

22.3 Implementation Case Example

Consider the implementation case described in table 39. This section shows how it can be worked out during the framework design definition stage. Its objective is to configure a proportional-derivative (PD) controller that process the attitude error on the spacecraft X-axis and generates a control torque for the same axis. It is assumed that the set point for the X-axis control is zero.

A controller of this kind trivially fits the controller model of figure 22.2 and can therefore be represented as an instance of class `Controller`.

According to the data exchange model of the intercomponent communication framelet, it will be assumed that the input and output for the controller come from a data pool. In a typical configuration, a single data pool is used to gather together all attitude-related data. It will contain, among others, a 3-vector representing the attitude measurement and a 3-vector representing the torque requests.

The PD control algorithm can be encapsulated in a control channel created and configured as shown in the implementation case of section 14.3.

With these premises, the configuration of the controller can be performed in pseudo-code as follows:

```
// Reference to the attitude data pool
AttitudeDataPool* attitudeDataPool;

// Reference to the PD control channel block
AbstractControlChannel* pd_block;

. . .

// Create the controller component
Controller* controller = new Controller();

// Retrieve the data items serving as inputs and outputs
// for the controller
AocsData* attErr = attitudeDataPool->getAttitudeError();
AocsData* torReq = attitudeDataPool->getTorqueRequest();
DataItemRead Xerror = attErr->getDataItemRead(1);
DataItemWrite Xtorque = torReq->getDataItemWrite(1);

// Configure the controller component
controller->linkMeasurementError(Xerror,1);
controller->linkActuatorRequest(Xtorque,1);
controller->setCompensatorTransferFunction(pd_block);
```

The last three statements assume that class `Controller` exposes methods `linkMeasurementError` and `linkActuatorRequest` to link the controller to its input (the attitude error) and to its output (the torque request). Method `setCompensatorTransferFunction` is instead used to load the transfer function from the measurement error to the actuator command. In general, class `Controller` should also expose methods to link the other input and output signals and to load the other transfer functions but these are not relevant to this implementation case.

The main point to note here is that, with the exception of the attitude data pool, all other components used in this example are in principle predefined by the framework. Hence the work required of the application developer to implement the desired controller simply reduces to performing a series of component composition operations. The objective of a framework is to reduce as much as possible of the application development to this type of operations.

Table 39. Implementation case example for controller framelet

Objective	Configure a PD controller for X-axis attitude control.
Description	The PD controller takes as input the attitude error on the X axis and generates a torque request to counteract it. The attitude reference for the controller is assumed to be zero.
Framelet	Controller Framelet Intercomponent Communication Framelet Sequential Data Processing Framelet
Constructs	`Controller` component `AocsDataPool` class `ControlChannel` class
Hot-Spots	Controllable hot-spot
Related ICs	IC of section 14.3
Implementation	- Create a `Controller` component - Link the controller input to the X-axis attitude error in a data pool - Link the controller output to the torque request in a data pool - Create a PD controller (see IC of section 14.3) - Load the PD component into the controller.

22.4 Reusability

Reusability of the controller concept within the AOCS domain is achieved by the usual device of separating the management of the controllers from the implementation of the control algorithms. The controller manager component only performs application-independent operations and can therefore be fully reused. Additionally, the controller component can be used to quickly build a closed-loop controller using control channels to encapsulate the control algorithms.

Implementation of control algorithms is a feature of all embedded control systems and the design solutions proposed by the controller framelets can be reused in other domains. Reuse at architectural and code level essentially involves reuse of the controller manager and controller class. Both are probably widely reusable because their logic is fairly general.

Table 40 summarizes the constructs exported by the framelet that have been identified or defined or identified in this chapter.

Table 40. Overview of controller framelet

Design Patterns	
Controller	Separate the management of the closed-loop controllers from their implementation.
Abstract Interfaces	
`Controllable`	Characterize a generic closed-loop controller.
Core Components	
`ControllerManager`	Component to manage the closed-loop controllers.
Default Components	
`Controller`	Configurable component implementing a classical closed-loop controller. The links to its data inputs and outputs and the control algorithms are settable parameters.
Hot-Spots	
Controllable	Applications must define the control algorithms for their closed-loop controllers by providing implementations for interface `Controllable`.

23 The Framework Instantiation Process

Frameworks exist to be instantiated and the final proof of their quality lies in the ease with which they allow applications within their target domain to be generated. To test a framework therefore essentially means to use it to construct a concrete application. The AOCS Framework was built as a research prototype and, as it stands, it certainly cannot be used to generate a piece of flight software. This chapter therefore cannot relate a "real" instantiation experience. It can, however, show what the main steps are in the instantiation process as they were followed to build simplified AOCS systems suitable for demonstration purposes. Additionally, it presents some results of measurements performed on such demonstration prototypes that give an idea of the overhead that use of framework technology introduces at application level.

23.1 Step-by-Step Instantiation

Instantiation of an application from a framework should normally be preceded by an analysis of the application requirements and by their matching to the framework functionalities to ascertain whether and to what extent the framework is suitable to help develop the application. If there is a high degree of matching and if the framework is a "good" framework offering a rich set of default components implementing most behaviour variants in the framework target domain, then the application development process largely takes the form of instantiating predefined framework components and combining them with each other. This section describes the main steps in this process. In a sense, it can be seen as an umbrella implementation case that describes the usage of the framework as a whole. Normally, the individual steps would be covered by dedicated implementation cases.

For each instantiation step a reference is given to the chapter where the relevant background material can be found. This book describes the framework at concept level only. The instantiation steps are described at a corresponding level of detail. Before they can function as a complete how-to guide to instantiating the framework, they need to be complemented with information that depends on a fuller definition of the framework. As they are, they are only useful to provide an indication of the complexity of the instantiation process.

The order in which the instantiation steps should be taken is not fixed. The order in which they are listed below is that used to instantiate a simplified AOCS application used as demonstrator for the framework.

A. Pasetti: Software Frameworks, LNCS 2231, pp. 271-278, 2002.
© Springer-Verlag Berlin Heidelberg 2002

1. *Instantiate and Configure the System Manager*: Create the system manager as the first application component because, in accordance with the system management pattern, all other components should register with it when they are created. The system manager is a core component and is predefined by the framework. See section 10.1.

2. *Instantiate and Configure the Event Repositories*: Analyse the application requirements to ascertain which kinds of events must be logged by the on-board software. For each category of events, the corresponding event repository must be instantiated and configured. In most cases, it will be possible to use the event categories and their associated event repositories predefined by the framework. If this is not the case, then a new event subclass together with its event repository must be defined by suitably subclassing `AocsEvent` and `EventRepository`. See section 13.1.

3. *Instantiate and Configure the Mission Mode Manager:* Most applications will need some way of relaying mode-related information from the ground or from the satellite central computer to the AOCS components. Hence, instantiate an AOCS Mission Mode Manager. This is a predefined framework component. See section 12.4.

4. *Instantiate and Configure the Data Pools:* Analyse the application requirements to identify which data are exchanged among the application components. These data must be stored in data pools so as to be available for access to all AOCS components. The content of the data pools is completely application-specific and therefore data pool components need to be constructed anew for each application. See section 13.4.

5. *Instantiate and Configure the Telemetry Stream:* The AOCS application will need to send telemetry data to a telemetry stream. Hence, construct a telemetry stream component. In many cases, it will be possible just to instantiate a default component predefined by the framework. See section 21.1.

6. *Instantiate and Configure the Telemetry Manager:* Instantiate and configure the telemetry manager component – predefined by the framework as a core component. If the format and content of the telemetry data is not constant during the mission, then the telemetry manager must be made mode-dependent and a mode manager must be defined for it. In some cases, it will be possible to use a predefined mode manager offered by the framework as a default component. See section 21.1.

7. *Instantiate and Configure the Controller Manager:* Instantiate and configure the controller manager component – predefined by the framework as a core component. If the control algorithms vary across mission phases (which is usually the case), then the controller manager must be made mode-dependent and a mode manager must be defined for it. In some cases, it will be possible to use a predefined mode manager offered by the framework as a default component. See section 22.1.

8. *Instantiate and Configure the Failure Recovery Manager:* Instantiate and configure the failure recovery manager component – predefined by the framework as a core component. If the failure recovery strategies vary across mission phases, then the failure recovery manager must be made mode-dependent and a mode manager must be defined for it. In some cases, it will be possible to use a predefined mode manager offered by the framework as a default component. See section 19.3.

9. *Instantiate, Configure and Load the Failure Recovery Strategies:* Analyse the application requirements to identify the recovery strategies required by the application and instantiate and configure the corresponding RecoveryStrategy components. In many cases, it will be possible to use default recovery strategies predefined by the framework. The recovery strategies thus defined are then loaded into the failure recovery manager. See section 19.2.

10. *Instantiate and Configure the Failure Detection Manager:* Instantiate and configure the failure detection manager component – predefined by the framework as a core component. If the failure detection checks vary across mission phases, then the failure detection manager must be made mode-dependent and a mode manager must be defined for it. In some cases, it will be possible to use a predefined mode manager offered by the framework as a default component. See section 18.3.

11. *Instantiate the AOCS Unit Components:* For each of the external units – sensors and actuators – required by the application, create a unit proxy component. The degree of variation across units is such that it is likely that this step will require construction of dedicated subclasses of AocsUnit. See section 15.2.

12. *Instantiate and Configure the AOCS Hardware Interface Components:* For each type of hardware interface to external units, instantiate and configure a hardware interface component. Since the number and type of such interfaces is limited, it will usually be possible to use predefined components offered by the framework as default components. See section 15.4.

13. *Configure the AOCS Unit Components*: Unit configuration is done in several steps. First, load the hardware interface components into the unit. Then link the unit component to its destination and source buffers in the data pool. Finally, load the conversion algorithms to go from raw data level to AOCS data level as control channel components. These latter components are probably application specific and will often have to be constructed from scratch. The data conversion algorithms however would typically be built in an Xmath/Matlab environment and therefore the corresponding components could be built using the autocoding facilities of these environments. The framework offers a wrapper to turn the autocoded routines into control channel components that are ready to be loaded into the units. See section 15.2.

14. *Instantiate the Unit Reconfiguration Managers*: In most cases, units are redundant. For each group of redundant units, instantiate a reconfiguration manager. The structure of redundancy groups and the reconfiguration logic is often the

same and therefore predefined components offered by the framework as default components can often be used for the reconfiguration managers. See section 16.2.

15.*Configure the Reconfiguration Managers*: Load the units into their respective configuration managers. From this point on, the reconfiguration managers are used in stead of the groups of units they represent. See section 16.2.

16.*Instantiate and Configure the Unit Triggers:* Analyse the application requirements to establish the times within the AOCS cycle when data transactions with the external units must be triggered. For each group of transactions that are to be performed at the same time, instantiate a unit trigger component. These component are predefined by the framework. If the type of units to be triggered or the order in which they must be triggered depends on the mission phase, then the unit triggers must be made mode-dependent and mode managers must be defined for them. In most cases, it will be possible to use predefined mode managers offered by the framework as default components. See section 15.3.

17.*Instantiate and Configure the Manoeuvre Manager:* Instantiate and configure the manoeuvre manager component – predefined by the framework as a core component. See section 17.2.

18.*Instantiate the Manoeuvre Components:* Analyse the application requirements to identify the manoeuvres that are required by the application. For each manoeuvre, instantiate the corresponding `Manoeuvre` subclass. In many cases, it will be possible to use predefined components offered by the framework as default components. The manoeuvre components are left unconfigured since they are configured only when they are used (for instance, a telecommand may be used to configure a manoeuvre component and then load it into the manoeuvre manager). See section 17.1.

19.*Instantiate and Configure the Telecommand Manager:* Instantiate and configure the telecommand manager component – predefined by the framework as a core component. See section 20.1.

20.*Instantiate the Telecommand Components:* Analyse the application requirements to identify the telecommands that are required by the application. For each type of telecommand, instantiate the corresponding `Telecommand` subclass. In many cases, it will be possible to use predefined components offered by the framework as default components. The telecommand components are left unconfigured since they are configured only when they are used (the configuration information is part of the data that are uplinked by the ground to the satellite). See section 20.1.

21.*Load Telemeterable Components:* Analyse the application requirements to determine which components must provide telemetry data in each mission phase. These components are then loaded into telemetry lists (one for each mission phase) that are in turn loaded into the telemetry mode manager. In most cases,

it will be possible to use the default implementations of method `writeToTe-`
`lemetry`. Where necessary this method can be overridden. See section 21.1.

22. *Load Failure Detection Components*: Analyse the application requirements to
determine which consistency checks must be performed in each mission phase
and which variables must be monitored in each mission phase. For each consistency check, provide an implementation of interface `ConsistencyCheck-`
`able`. In many cases, it will be possible to use the default implementation of
this interface built into components predefined by the framework. For each variable to be monitored, instantiate and configure a `MonitoringCheck` component. The consistency checkable and the monitoring check components are
then loaded into lists that are in turn loaded into the failure detection mode manager. See sections 18.2.1 and 18.2.2

23. *Load Unit Components in the Unit Triggers:* Load the components representing the external units into the mode managers of the unit triggers. See section
15.3.

24. *Load the Controller Components:* Analyse the application requirements to determine which control algorithms are used in each mission phase. For each
control algorithm, a control channel component must be defined. If the control
algorithms were synthesized within an Xmath/Matlab environment, their autocoded versions can be directly inserted into the wrappers provided by the framework as default components. The control channel components are then loaded into `Controller` components which are finally loaded into the controller
mode manager. See section 22.2.

25. *Instantiate, Configure and Load the Recovery Actions:* Analyse the application
requirements to determine the recovery actions to be associated to each failure
detection check prescribed by the application (recall that the framework prescribes that to each failure detection check a recovery action should be associated). For each recovery action thus identified, instantiate and configure the
corresponding `RecoveryAction` subclass. In many cases, it will be possible
to use predefined components offered by the framework as default components.
The recovery actions must then be loaded into the components that perform the
failure detection checks. Since most components in the framework perform
such checks, this step is performed in parallel to the previous ones. See section
19.1.

26. *Perform Configuration Check:* through the system manager component, perform a configuration check to verify that all instantiated components are correctly configured. See section 10.3.

The main point to note about the instantiation procedure outlined above is that,
in most cases, the actions that are required of the application developer are instantiation and configuration of existing components. Configuration usually means
setting parameter values (e.g. setting the maximum number of events that can be
held by an event repository) and plugging components with each other (e.g. loa-

ding a recovery strategy into the failure recovery manager). The main exceptions are the definition of the data pools, the definition of the proxies for the external units and the definition of the components encapsulating control algorithms. The latter, however, can often be done through an autocoding tool in an Xmath/Matlab environment.

The complexity of instantiating and configuring a large number of components should not be underestimated. A typical component may require five or six configuration actions most of which set up a link with other components. This results in a complex web of component interactions that can be difficult to master both conceptually and syntactically. However, component configuration is definitely easier than component design. Above all, the configuration process within a framework context follows rules (often embodied in implementation cases) which are part of the framework itself. This is because the framework predefines the component interactions as well as the component themselves. Even more importantly, most instantiation steps could in principle be automatized. It is possible to imagine an environment with a graphical user interface where users can pull down components from palettes of predefined components, instantiate and link them with each other graphically and set their configuration parameters through wizards. The instantiation code is then automatically created in the background. Such environments already exist for creating graphical applications and there is no reason why they should not exist for other domains where the framework approach is adopted. This type of environments is behind the vision outlined in the introduction and schematically represented in figure 1.3. We hope that the discussion in this chapter has shown how a framework brings their concrete realization closer.

23.2 Framework Overheads

Frameworks make the instantiation of an application within their domain easier. The first and most visible price to be paid for this greater ease of development is the investment that is required to create the framework. There is then a – rather smaller – price to be paid in terms of reduced performance which is due to the overheads that are introduced at application level by the framework. In the case of the AOCS Framework, its structure as a domain-specific extension of the operating system (see section 8.6), made it possible to quantify them. Note that the emphasis here is on the overheads that accrue to the use of the framework itself and not to those due to the use of object-orientation. The latter are well-known and to some extent have already been quantified elsewhere [34, 48].

The framework overhead measurements were made using the C++ version of the framework instantiated on an ERC32 processor with the RTEMS operating system. The ERC32 processor [37] is a space-qualified version of a SPARC V7 processor. For the experiments described here it ran at 14 MHz. The comparatively low frequency is typical of hardware used in space where the radiation intensive environment makes operation at high frequency error-prone. RTEMS [85] is a public domain operating system for embedded processors.

Consider first the timing overheads. The framework leads to an application that is constructed as a bundle of functionality managers that are activated in sequence and that perform standard operations upon a configurable list of clients. Thus, for instance, the telemetry manager performs the telemetry operations upon a list of client components of type `Telemeterable`. This list is loaded into the telemetry manager at configuration time. The timing overheads that are intrinsic to the framework can be measurement by instantiating a dummy application that includes all the functionality managers defined by the framework but leaves them "empty". Thus, for instance, the telemetry manager is created but its telemetry lists are not loaded with any telemeterable components. The result is an AOCS application where the functionality managers are activated and run normally but do not have any clients upon which to operate. This represents the smallest and fastest executable application that can be instantiated from the framework. The time it takes to executes one full cycle – to activate all the functionality managers and return – can be taken as a measure of the timing overhead introduced by the framework. Measurements with the above processor configuration indicate that one cycle of such an empty AOCS application requires 0.2 ms of processor time. This figure should be set against typical AOCS cycles of duration of 50 ms or more. The framework-induced CPU overhead is therefore less than 1% and can be considered as negligible.

Consider now the memory overhead. In order to estimate it, full executable modules were loaded into an ERC32 simulator and the memory usage reported by the simulator was then recorded. Two different executables were used for this purpose:

- `EmptyMemoryTest`: this executable is built from a dummy main program. Its memory requirements essentially correspond to the memory requirements of the C++ run-time systems (excluding the RTEMS components).
- `MemoryTestOnlyFunctMan`: this executable links together the framework functionality managers and no other plug-in components. It therefore represents the core that has to be included by any AOCS generated from the framework.

The memory requirements of these modules as reported by the ERC32 simulator are shown in table 41. The framework functionality managers, after subtraction of the "empty memory test" requirements, represent in a sense the "overhead" introduced by the framework. From the figure in the table, this overhead can be computed as: (74-30)+(19-2)+2=63 Kbytes. Since a typical AOCS application might occupy around 400-600 Kbytes, the framework memory overhead is around 10-15%. This is not negligible but in view of the expansion in memory availability in space applications it is definitely an acceptable price to pay for the convenience of the framework approach.

Table 41. Results of memory measurements

	EmptyMemoryTest	MemoryTestOnlyFunctMan
.text	30 Kbytes	74 Kbytes
.data	2 Kbytes	19 Kbytes
.bss	0	2 Kbytes

Appendix

This appendix presents three summary tables for the AOCS Framework which list its domain abstractions, its design patterns and its hot-spots as they were identified or defined in this book. Cross-references are provided to the sections in this book where each of the items in the lists is described in greater detail.

The domain dictionary (see section 6.3.2) defines the domain abstractions. The domain abstractions capture the commonalities of the AOCS applications in the framework domain. They describe concepts and functional features that span application boundaries and are found in all domain applications. Domain abstractions provide the vocabulary to describe the AOCS Framework architecture. The domain dictionary is shown in table 42.

Table 43 lists the framework-level design patterns defined in this book together with a brief description of the design problem they address and, where appropriate, of the patterns of which they are a specialization or which inspired their design. Mention is also made of patterns that introduce recursion. This is important in the AOCS domain because recursion makes static predictability of timing property impossible without semantic information (see section 9.10).

Finally, table 44 lists the framework-level hot-spots of the AOCS Framework.

Table 42. Domain dictionary for the AOCS Framework

Dictionary Entry	Definition
Abstract Control Channel	Encapsulation of a generic multi-input-multi-output transfer function. See section 14.2.
AOCS Data	Data that are produced or consumed on a periodic basis by AOCS components (also known as *cyclical data*). See section 13.3.
AOCS Event	An object that encapsulate a description of an asynchronous event. See section 13.1.
AOCS Object	An object that is instantiated, directly or indirectly, from class `AocsObject` and that has access to the following basic services: time recovery, failure reporting, and configuration error reporting; and has the following properties: resettability, configurability, and telemeterability. See section 9.7.
Bound Property	A property that is subject to monitoring through change notification. See section 11.3.2.
AOCS Unit	Any device (normally a sensor or an actuator) that is external to the AOCS computer but internal to the AOCS system. See section 14.

A. Pasetti: Software Frameworks, LNCS 2231, pp. 279-284, 2002.
© Springer-Verlag Berlin Heidelberg 2002

Table 42. (cont.)

Change Object	Encapsulation of a specific time profile for a property value. See section 11.2.
Configurable	Property of a component that can clear its internal configuration and that can check whether it is correctly configured. See section 10.3.
Consistency Check	Failure detection test where a component is asked to check its own internal consistency. See section 18.2.1.
Control Block	An abstract control channel that cannot be further decomposed into lower level control blocks. See section 14.1.
Control Channel	Same as Abstract Control Channel above.
Control Super Block	Container for a chain of interconnected control channels. See section 14.1.
Controller	Encapsulation of a generic closed-loop controller. See section 22.2.
Cyclical Data	Same as AOCS Data above.
Data Pool	Shared memory area where AOCS data are stored. See section 13.4.
Event Repository	Shared memory area where AOCS events are stored. See section 13.1.
Failure	Any fault that is detected during the normal operation of a component and that affects its ability to perform its allotted task. Failures would normally occur only during normal operation. See section 18.2.1.
Fictitious AOCS Unit	Component that looks to its clients like a proxy for an external unit. See section 15.5.
Manoeuvre	Sequence of actions to be performed by the AOCS at specified times to achieve a specified goal. See section 17.1.
Mode Change Action	Actions to be performed when a mode-dependent component changes mode. See section 12.2.
Monitor	A component that performs a monitoring action. Also called a monitoring component. See section 11.3.
Monitored Property	A property that is subject to a monitoring action by a monitor. See section 11.3.
Monitoring Action	Observation of a change over time in the value of a property. See section 11.3.
Monitoring Check	Failure detection test where it is verified that the value of a property follows a pre-specified change. See section 18.2.2.
Operational Mode	Attribute of a component whose behaviour depends on operational conditions. See chapter 12.
Property	Attribute of a component that describes one aspect of its behaviour or of its internal state and that is externally accessible. See section 11.1.
Property Object	Object that encapsulates a property. See section 11.1.
Reconfiguration	Switch between different independent implementation of the same functionality. See section 16.1.

Table 42. (cont.)

Reconfiguration Group	Set of components that together offer a redundant functionality. See section 16.1.
Recovery Action	Set of actions to be taken in response to a single failure report. See section 19.1.
Recovery Strategy	Set of actions to be taken in response to a set of failure reports. See section 19.1.
Redundant Functionality	Functionality for which two or more independent implementations are provided. See section 16.1.
Redundant Object	Member of a reconfiguration group. See section 16.1.
Resettable	Property of a component that can bring its internal state to some initial default value. See section 10.2.
Sequential Data Processing Chain	Sequence of independent processing stages through which AOCS data are passed. See chapter 14.
Strategy	Encapsulation of a mode-dependent behaviour: to each operational mode there corresponds a strategy. See section 19.2.
System Management Function	A function that is performed systematically on all AOCS components present in the AOCS software at a given time. The system manager is the component responsible for performing system management functions. See chapter 10.
Telecommand	Encapsulation of an action to be performed upon the AOCS software originating in the ground station. See chapter 20.
Telecommand Loading	Process whereby telecommands are loaded from an external interface ultimately communicating with the ground station into the AOCS application software. See section 20.3.
Telecommand Transaction	A sequence of telecommands that are executed as a single transaction and whose actions, in case of partial or complete failure, can be reversed to leave the AOCS in the same state in which it was before the transaction was started. See section 20.2.
Telemeterable	Property of a component whose state can potentially be included in the telemetry stream. See section 21.1.
Telemetry Frame	Set of telemetry data sent to the telemetry stream in one single telemetry cycle. See chapter 21.
Telemetry Stream	Data sink to which telemetry data are written representing the channel through which telemetry data are sent to the ground. See section 21.1.
Transaction Telecommand	One of the telecommands making up a telecommand transaction. See section 20.2.

Table 43. Framework-level design patterns for the AOCS Framework

Design Pattern	Summary Description
Control Channel D.P.	Pattern to allow control channels to be treated uniformly regardless of whether they represent control blocks or chains of interconnected and nested control blocks. Based on composite patter. This pattern introduces recursion. See section 14.2
Controller D.P.	Pattern to separate the management of closed-loop controllers from their implementation. Based on the manager meta-pattern. See section 22.1.
Direct Property Monitoring D.P.	Pattern to directly monitor a property by accessing the property value or the property object through getter methods exposed by the property owner. Inspired by the JavaBeans architecture. See section 11.3.1.
Failure Detection D.P.	Pattern to separate the management of failure detection tests from their implementation. Based on the manager meta-pattern. See section 18.3.
Failure Recovery D.P.	Pattern to separate the management of failure recovery from the implementation of failure recovery actions. Based on the manager meta-pattern and on the chain of responsibility pattern. This pattern introduces recursion. See section 19.3.
Fictitious Unit D.P.	Pattern to address the problem of combining components that process unit data without impacting the final users of the unit data. This pattern introduces recursion. See section 15.5.1.
Manoeuvre D.P.	Pattern to separate the management of manoeuvres from their implementation. Based on the manager meta-pattern. See section 17.2.
Mode Management D.P.	Pattern to endow components with mode-dependent behaviour while separating the implementation of the mode-dependent strategies from the implementation of the logic governing mode switches. See section 12.1.
Monitoring through Change Notification D.P.	Pattern to monitor a property by registering with the property owner and asking it to notify the monitor when a change of a certain type (as encapsulated by a change object) has occurred. Inspired by the JavaBeans architecture. See section 11.3.2.
Reconfiguration Management D.P.	Pattern to address the problem of separating the management of a reconfiguration group from the provision of the reconfigurable functionality. See section 16.2.
Shared Data D.P.	Pattern to address the problem of allowing components to share access to data that are generated synchronously. See section 13.2.
Shared Event D.P.	Pattern to address the problem of allowing components to share access to event that are generated asynchronously. See section 13.1.

Table 43. (cont.)

System Management D.P.	Pattern to address the problem of systematically performing the same set of operations on a target set of components. Based on the manager meta-pattern. See section 10.1.
Telecommand Management D.P.	Pattern to address the problem of separating the management of telecommands from their implementation. Based on the manager meta-pattern. See section 20.1.
Telemetry Management D.P.	Pattern to address the problem of making the management of telemetry data independent of the layout and content of telemetry frames. Based on the manager meta-pattern. See section 21.1.

Table 44. Framework-level hot-spots for the AOCS Framework

Hot-Spot	Summary Description
AOCS Event H.S.	Applications must subclass `AocsEvent` to define their own event classes. See section 13.1.
AOCS Data H.S.	Applications must subclass `AocsData` to define their own data types. See section 13.2.
AOCS Unit H.S.	Applications must subclass `AocsUnit` to encapsulate interactions with external units. See section 15.2.
AOCS Unit Hardware H.S.	Applications must implement interface `AocsUnitHardware` to encapsulate the interaction with the low-level hardware that controls data exchanges with external units. See section 15.2.
Change Object H.S.	Applications must implement interface `ChangeObject` to encapsulate the change profiles in which they are interested. See section 11.2.
Configurable H.S.	AOCS objects must provide an implementation of interface `Configurable` to clear their internal configuration and to check that they have been configured. See section 10.3.
Consistency Checkable H.S.	AOCS objects must provide an implementation of interface `ConsistencyCheckable` to define their application-specific consistency checks. See section 18.2.1.
Control Channel H.S.	Applications must provide the definition of propagation algorithms for the control channels they use. See section 14.2.
Controllable H.S.	Applications must provide the definition of their closed-loop controllers. See section 22.1.
Data Pool H.S.	Applications must define the data pools where application data are stored. A data pool is created by deriving a class from `AocsDataPool`. The derived class contains AOCS data objects that are logically related. See section 13.4.
Fictitious Unit H.S.	Applications must implement interface `AocsUnitFunctional` to encapsulate the processing of unit data. See section 15.5.

Table 44. (cont.)

Manoeuvre H.S.	Applications must define the application-dependent manoeuvres by subclassing class `Manoeuvre`. See section 17.1.
Mode Change Action H.S.	Applications must define the application-specific actions to be associated to mode changes in mode-dependent components. See section 12.2.
Mode Implementer H.S.	Applications must define the application-dependent strategy implementations for mode-dependent components. See section 12.1.
Mode Manager H.S.	Applications must define the application-dependent mode switching logic for mode-dependent components. See section 12.1.
Monitoring Check H.S.	Applications must define the properties that need to be monitored during failure detection tests and the type of change in their value that determines a failure. See section 18.2.2.
Resettable H.S.	AOCS objects must provide an implementation of interface `Resettable` to reset their internal state. See section 10.2.
Telecommand H.S.	Applications must provide their own application-specific telecommands by subclassing the base class `Telecommand`. See section 20.1.
Telecommand Loader H.S.	Application must provide their own application-specific telecommand loader by implementing the abstract interface `TelecommandLoader`. See section 20.3.
Telemetry H.S.	Applications must provide their own application-specific implementation of interface `Telemeterable` defining the content of the telemetry images of their components. See section 21.1.
Telemetry Stream H.S.	Application must provide an application-specific implementation of interface `TelemetryStream` defining their telemetry stream. See section 21.1.
Unit Trigger H.S.	Application must define the order in which units are triggered and the type of triggering action. See section 15.3

References

1. Adams M, Grib T, (1999) A Component Based, Event-Driven Framework for Rapid Prototyping of Real-Time Avionics Systems. Proceedings of the 18-th Digital Avionics and Space Conference, Phoenix (AZ), USA, 1999, paper 9.C.5
2. Ardis M et al (2000) Domain Engineered Configuration Control. In: Donohoe P (ed) Software Product Lines – Experience and Research Directions. Kluwer Academic Publisher, pp 479-494
3. Aksit M, Tekinerdogan B, Marcelloni F (1999) Deriving Frameworks from Domain Knowledge. In: Fayad M, Schmidt D, Johnson R (eds.) Building Application Frameworks – Object Oriented Foundations of Framework Design. Wiley Computer Publishing, pp 169-98
4. America P, et al (2000) CoPAM: A Component-Oriented Platform Architecting Method Family for Product Family Engineering. In: Donohoe P (ed) Software Product Lines – Experience and Research Directions. Kluwer Academic Publisher, pp 167-180
5. Atkinson C, Bayer J, Muthig D (2000) Component-Based Product Line Development: The KobrA Approach. In: Donohoe P (ed) Software Product Lines – Experience and Research Directions. Kluwer Academic Publisher, pp 289-310
6. Awad M, Kuusela J, Ziegler J (1996) Object Oriented Technology for Real Time Systems. Prentice Hall
7. Barr M (1999) Programming Embedded Systems. O'Reilly
8. Barnes J (1997) Ada 95 rationale : the language, the standard libraries. Heidelberg: Springer Verlag
9. Bass L, Clements P, Kazman R (1998) Software Architecture in Practice. Addison Wesley Longman
10. Batory et al (1995) Creating Reference Architectures: An Example from Avionics. Symposium on Software Reusability, Seattle, Washington, USA. Available from this web site: ftp://ftp.cs.utexas.edu/pub/predator/adage-arch.pdf
11. Batory D et al (1995) The ADAGE Avionics Reference Architectures. AIAA Computing in Aerospace Conference, S. Antonio (CA).
12. Batory D, Geraci B (1996) Validating Component Composition in Software System Generators. Proceedings of The Fourth International Conference on Software Reuse (ICSR '96)
13. Batory D, Cardone R, Smaragdakis Y (2000) Object-Oriented Frameworks and Product Lines, In: Donohoe P (ed) Software Product Lines – Experience and Research Directions. Kluwer Academic Publisher, pp 227-248
14. Beck K (1999) Extreme Programming Explained. Addison-Wesley

A. Pasetti: Software Frameworks, LNCS 2231, pp. 285-290, 2002.
© Springer-Verlag Berlin Heidelberg 2002

15.Blum M, Kannan S (1989) Designing Programs that check their Work. Proceedings of the ACM 21st Annual Symposium on the Theory of Computing, May
1989

16.Bockle G (2000) Model-Based Requirements Engineering for Product Lines.
In: Donohoe P (ed) Software Product Lines – Experience and Research Directions. Kluwer Academic Publisher, pp 193-204

17.Booch G (1991) Object-Oriented Design with Applications. Benjamin-
Cummings

18.Booch G, Rumbaugh J, Jacobson I (1996) Unified Modelling Language. Ratio-
nal Software

19.Bosch J, et al (1999) Framework Problems and Experiences. In: Fayad M,
Schmidt D, Johnson R (eds.) Building Application Frameworks – Object Oriented Foundations of Framework Design. Wiley Computer Publishing, pp 55-
82

20.Bosch J (2000) Design and Use of Software Architectures. Addison-Wesley

21.Brooks F (1995) The Mythical Man-Month. Addison-Wesley

22.Brown T et al (to be published) A Reusable and platform-independent framework for distributed control systems. Proceedings of the 20-th Digital Avionics
Systems Conference, Daytona Beach, FL (USA), Oct. 2001

23.BSSC - ESA Board for Software Standardization and Control (1998) Guide to
Applying the ESA Software Standards to Projects Using Object-Oriented Methods. European Space Agency

24.Buschmann F et al (1996) A System of Patterns – Pattern Oriented Software
Architectures. Wiley

25.Clemens P, Northrop L (2001) A Framework for Software Product Line Practice. Software Engineering Institute, Carnegie Mellon University, Available
from Internet Web Site: www.sei.cmu.edu/activities/plp/framework.html

26.Coad P, Yourdon E (1991) Object-Oriented Design. Prentice Hall

27.Codenie W et al (1997), From Custom Applications to Domain-Specific Frameworks, Communications of the ACM, Vol. 40, N. 10, page 71-77

28.Collins D (1995) Designing Object-Oriented User Interfaces. The Benja-
min/Cummings Publishing Company

29.Coplien J, Hoffman D, Weiss D (1998) Commonality and Variability in Software Engineering. IEEE Software, Dec. 1998, pag. 37-45

30.Coriat M, Jourdan J, Boisbourdin F (2000) The SPLIT Method. In: Donohoe P
(ed) Software Product Lines – Experience and Research Directions. Kluwer
Academic Publisher, pp 147-166

31.Dager J (2000) Cummins' Experience in Developing a Software Product Line
Architecture for Real-Time Embedded Diesel Engine Control. In: Donohoe P
(ed) Software Product Lines – Experience and Research Directions. Kluwer
Academic Publisher, pp 23-46

32.Dellinger W, Salada M, Shapiro H (1999) Application of Matlab/Simulink to
Guidance and Control Flight-Code Design. Proceedings of the 22nd AAS Guidance and Control Conference, Feb. 1999, Breckenridge, Colorado, USA

33.Doerr B, Sharp D, Freeing Product Line Architectures from Execution Dependencies, In: Donohoe P (ed) Software Product Lines – Experience and Research Directions. Kluwer Academic Publisher, pp 313-330

34.Driesel K (1996) The Direct Cost of Virtual Function Calls in C++. ACM Sigplan Notices, Vol. 31, n. 10, pp 306-316

35.ECSS Secretariat (1999) Space Engineering Standard – Software ECSS-E40. Issue A, http://www.estec.esa.nl/ecss/

36.Englander R (1997) Developing Java Beans. O'Reilly

37.ERC32 Home Page, http://www.estec.esa.nl/wsmwww/erc32/erc32.html

38.ERC32 Free Software, http://www.estec.esa.nl/wsmwww/erc32/freesoft.html

39.Fayad M, Schmidt D, R. Johnson R (1999) Application Frameworks. In: Fayad M, Schmidt D, Johnson R (eds.) Building Application Frameworks – Object Oriented Foundations of Framework Design. Wiley Computer Publishing, pp 3-28

40.Fontoura M (1999) A Systematic Approach to Framework Development. PhD Thesis, Computer Science Department, Pontificial Catholic University of Rio de Janeiro

41.Fontoura M, Pree W, Rumpe B (2001) The UML Profile for Object Frameworks. Addison Wesley

42.Froelich G, et al (1999) Reusing Hooks. In: Fayad M, Schmidt D, Johnson R (eds.) Building Application Frameworks – Object Oriented Foundations of Framework Design. Wiley Computer Publishing, pp 219-236

43.Gamma E (1991) Objektorientierte Software-Entwicklung am Beispiel von ET++: Design-Muster, Klassenbibliothek, Werkzeuge. Doctoral Thesis, University of Zürich (published by Springer Verlag, 1992)

44.Gamma E, et al. (1995) Design Patterns – Elements of Reusable Object Oriented Software. Addison-Wesley, Reading, Massachusetts

45.Gill C, Levine D, Schmidt D (2001) The Design and Performance of a Real-Time CORBA Scheduling Service. Real-Time Systems: the International Journal of Time-Critical Computing Systems, Vol. 20 No. 2, pp 117-154

46.Greham R, Moote R, Cyliax I (1998) Real-Time Programming: A Guide to 32-bit Embedded Development. Addison-Wesley

47.Hein A et al (2000) Applying Feature Models in Industrial Settings. In: Donohoe P (ed) Software Product Lines – Experience and Research Directions. Kluwer Academic Publisher, pp 47-70

48.Herity D (1998) C++ In Embedded Systems: Myth and Reality. Embedded Systems Programming, Feb. 1998

49.Hood Technical Group (1995) Hood Reference Manual. Release 4, June 1995.

50. IEEE Standard 1471-2000, IEEE Recommended Practice for Description of Software Intensive Systems

51.Jacobson E, Nowack P (1999) Frameworks and Patterns: Architectural Abstractions. In: Fayad M, Schmidt D, Johnson R (eds.) Building Application Frameworks – Object Oriented Foundations of Framework Design. Wiley Computer Publishing, pp 29-54

52. Jolin A (1999) Usability and Framework Design. In: Fayad M, Schmidt D, Johnson R (eds.) Building Application Frameworks – Object Oriented Foundations of Framework Design. Wiley Computer Publishing, pp 153-162

53. Johnson R (1997) Frameworks=Components+Patterns. Communications of the ACM, Vol. 40, N. 10, pp39-42

54. Jones S (1999) A Framework Recipe. In: Fayad M, Schmidt D, Johnson R (eds.) Building Application Frameworks – Object Oriented Foundations of Framework Design. Wiley Computer Publishing, pp 237-265

55. Kang K et al (1998) FORM: A Feature-Oriented Reuse Method with Domain-Specific Reference Architectures. Annals of Software Engineering, Vol. 5, pag. 143-168

56. Kang K et al (1999) Feature-Oriented Engineering of PBX Software for Adaptability and Reusability. Software Practice and Experience, Vol 29(10), pag. 875-896

57. Kiczales G et al (1997) Aspect-Oriented Programming. In: Aksit M, Matsuoka S (Eds.) Proceedings ECOOP'97, LNCS 1241, Springer-Verlag, Jyvaskyla, Finland, June 1997, pp. 220-242

58. Kishi T, Noda N (2000) Aspect-Oriented Analysis for Product Line Architecture. In: Donohoe P (ed) Software Product Lines – Experience and Research Directions. Kluwer Academic Publisher, pp 135-146

59. Kottman M, Qiu X, Schaufelberger W (2000) Simulation and Computer Aided Control Systems Design Using Object Orientation. Hochschulverlag AG, ETH-Zurich

60. Krasner G, Pope S (1988) A Cookbook for Using the Model-View-Controller User Interface Paradigm in Smalltalk-80. Journal of Object-Oriented Programming, 1(3), 1988, pp 26-49

61. Lafontaine J et al (1999) Development of the Proba Attitude Control and Navigation Software. Proceedings of the 4th ESA International Conference on Spacecraft Guidance, Navigation and Control Systems, Noordwijk, The Netherlands, pp 427-442

62. Larson W, Wertz J (eds) (1997) Space Mission Analysis and Design. Space Technology Series, Kluwer Academic Publishers

63. Lee K et al (2000) Domain-Oriented Engineering of Elevator Control Software. In: Donohoe P (ed) Software Product Lines – Experience and Research Directions. Kluwer Academic Publisher, pp 3-22

64. Lewis T, et al (1996) Object-Oriented Application Frameworks. Manning Publications/Prentice Hall

65. Matlab Home Page, http://www.mathworks.com/

66. Mattsson M, Bosch J (1999) Composition Problems, Causes, and Solutions. In: Fayad M, Schmidt D, Johnson R (eds.) Building Application Frameworks – Object Oriented Foundations of Framework Design. Wiley Computer Publishing, pp 467-487

67. Meekel J et al (1997) From Domain Models to Architecture Frameworks. Software Engineering Notes, Vol 22, N. 3

68.Miller G et al (1999) Capturing Framework Requirements. In: Fayad M, Schmidt D, Johnson R (eds.) Building Application Frameworks – Object Oriented Foundations of Framework Design. Wiley Computer Publishing, pp 309-324

69.Morelli G (2001) Personal communication to the author

70.Murphy N (1998) Introduction to CORBA for Embedded Systems. Embedded Systems Programming Magazine, Oct. 98, p. 60-73

71.OBOSS Home Page, http://spd-web.terma.com/Projects/OBOSS/Home_Page/

72.Pasetti A, Pree W (1999) A Component Framework for Satellite On-Board Software. Proceedings of the 18-th Digital Avionics Systems Conference, St. Louis (USA), Oct. 99, paper 7.C.1

73.Pasetti A, Pree W (1999) The Component Software Challenge for Real-Time Systems. Proceedings of the Real-Time Mission Critical Systems Workshop (held in conjunction with 20th IEEE RTSS), 30 Nov - 1 Dec 1999, Scottsdale, AZ (USA)

74.Pasetti A, Pree W (2000) Two Novel Concepts for Systematic Product Line Development. In: Donohoe P (ed) Software Product Lines – Experience and Research Directions. Kluwer Academic Publisher, pp 249-270

75. Pasetti A, et al (to be published) An Object-Oriented Component-Based Framework for On-Board Software. Proceedings of the Data Systems In Aerospace Conference, Nice, May 2001

76. Pree W (1994) Meta Patterns – A Means of Capturing the Essential of Reusable Object Oriented Design. Proceedings ECOOP'94. In: Tokoro M, Pareschi R (Eds.), LNCS 821, Springer-Verlag, Bologna, Italy, July 1994, pp. 150-162.

77.Pree W (1995) Design Patterns for Object Oriented Software Development. Addison-Wesley, Reading MA

78.Pree W, Koskimies K (1999) Framelets – Small is Beautiful. In: Fayad M, Schmidt D, Johnson R (eds.) Building Application Frameworks – Object Oriented Foundations of Framework Design. Wiley Computer Publishing, pp 411-414

79.Pree W, Koskimies K (1999) Rearchitecturing Legacy Systems—Concepts & Case Study WICSA '99: First Working IFIP Conference on Software Architecture, San Antonio, Texas, 22-24 Feb. 1999

80.Pree W (1999) Hot-Spot Driven Development. In: Fayad M, Schmidt D, Johnson R (eds.) Building Application Frameworks – Object Oriented Foundations of Framework Design. Wiley Computer Publishing, pp 379-394

81.Pree W, Pasetti A, Brown T (2001) Hints and Guidelines for the Framework Development and Adaptation Process. In: Fontoura M, Pree W, Rumpe B, The UML-F Profile for Framework Architectures. Pearson Education/Addison-Wesley

82.Pronk B (2000) An Interface-Based Platform Approach. In: Donohoe P (ed) Software Product Lines – Experience and Research Directions. Kluwer Academic Publisher, pp 331-352

83.Reinhart T, Boettcher C, Tomashefsky S (1999) Self-Checking Software: Improving the Quality of Mission-Critical Systems. Proceedings of the 18-th Digital Avionics and Space Conference, St. Louis (MO), USA, Oct. 1999, paper 2.D.4

84. Reinhart T et al (2000) Self-Checking Software for Information Assurance in the 21st Century. Proceedings of the 19-th Digital Avionics and Space Conference, Philadelphia (PA), USA, Oct. 2000, paper 1.B.5

85. RTEMS Home Page, http://www.rtems.com

86. Real-Time for Java Expert Group (to be published) The Real-Time Specification for Java. Addison-Wesley. Draft available from: http://www.rtj.org/

87. Real-Time Platform Special Interest Group (PSIG), Internet home page address: http://www.omg.org/realtime/

88. Rumbaugh J, et al (1991) Object-Oriented Modelling and Design. Prentice-Hall

89. Savigni A, Tisato F (2000) Real-Time Programming in the Large. Proceedings of the 3rd IEEE International Symposium on Object-Oriented Real-Time Distributed Computing, Newport Beach (CA), USA, Mar. 2000, pp 352-355

90. Schmid A (1997) Systematic Framework Design by generalization. Communications of the ACM, Vol. 40, n. 10, pp 48-51

91. Schmid K (2000) Scoping Software Product Lines. In: Donohoe P (ed) Software Product Lines – Experience and Research Directions. Kluwer Academic Publisher, pp 513-532

92. Shaw M, Garlan D (1996) Software Architectures – Perspectives on an Emerging Discipline. Prentice Hall

93. Sharp D (2000) Component-Based Product Line Development of Avionics Software. In: Donohoe P (ed) Software Product Lines – Experience and Research Directions. Kluwer Academic Publisher, pp 353-370

94. Shlaer S, Mellor S (1988) Object Oriented Systems Analysis – Modelling the World in Data. Yourdon Press

95. Simon D (1999) An Embedded Software Primer. Addison-Wesley

96. Society of Automotive Engineers (1996) Generic Open Architecture (GOA) Framework. SAE Document AS4893

97. Soundarajan N (1999) Understanding Frameworks. In: Fayad M, Schmidt D, Johnson R (eds.) Building Application Frameworks – Object Oriented Foundations of Framework Design. Wiley Computer Publishing, pp 289-308

98. Szyperski C (1998) Component Software. Addison Wesley Longman Limited, Harrow (UK)

99. Toft P, Coleman D, Otha J (2000) A Cooperative Model for Cross-Divisional Product Development. In: Donohoe P (ed) Software Product Lines – Experience and Research Directions. Kluwer Academic Publisher, pp 111-132

100. Xmath Home Page, http://www.wrs.com/products/html/xmath.html

101. Wertz J (1995) Spacecraft Attitude Determination and Control. Kluwer Academic Publishers

102. Wills L et al (2000) An Open Control Platform for Reconfigurable, Distributed, Hierarchical Control Systems. Proceedings of the 19-th Digital Avionics and Space Conference, Philadelphia (PA), USA, paper 4-D-2

Index

Lecture Notes in Computer Science

For information about Vols. 1–2185
please contact your bookseller or Springer-Verlag